PLANNED SPRAWL

Volume 38, Sage Library of Social Research

"The whole of urban planning must be understood as a sector of the publicity propaganda effort of our society. It's the organization of participation in something in which it is impossible to participate."

—Theses on Unitary Urbanism

Planned Sprawl

Private and Public Interests in Suburbia

MARK GOTTDIENER

Preface by GERALD D. SUTTLES

VOLUME 38
SAGE LIBRARY OF
SOCIAL RESEARCH

SAGE PUBLICATIONS   Beverly Hills   London

For information address:

SAGE PUBLICATIONS, INC.
275 South Beverly Drive
Beverly Hills, California 90212

SAGE PUBLICATIONS LTD
St George's House / 44 Hatton Garden
London EC1N 8ER

Printed in the United States of America

Library of Congress Cataloging in Publication Data

Gottdiener, Mark.
Planned sprawl.

(Sage library of social research ; v. 38)
Bibliography: p. 183
1. Suburbs–United States–Case studies. 2. Regional planning–New York (State)–Suffolk Co. I. Title.
HT351.G66 309.2'5'0973 76-53960
ISBN 0-8039-0593-9
ISBN 0-8039-0594-7 pbk.

FIRST PRINTING

CONTENTS

for Jennifer

PREFACE

America has often been characterized as the "planless society." Although there is a good deal of truth in this image, the nation's history is dotted with efforts to bring urban growth under control. From the City Beautiful movement to the urban agenda of the Kennedy administration, Americans have persistently struggled against the tide of urban sprawl and obsolescence. Yet despite vigorous efforts, the accomplishments have been rather modest. It would be wrong, however, to say that such efforts were inconsequential. Rather the consequences have been unanticipated and frequently unpopular. Recent examples easily come to mind. The construction of public housing did not especially increase our housing stock and has probably added to high levels of housing abandonment and a growing uncertainty among remaining urban home owners. Federal housing subsidies and tax relief in all likelihood have increased suburbanward movement at the expense of our central cities.

There can be little doubt that the federal urban agenda of the 1960s is now in disarray and that it has been more or less abandoned. In the aftermath of this experience, two general reactions can be singled out in both popular debate and the technical literature. First, there are those who seem simply to be disenchanted with the aspiration of urban planning and to have a vague feeling that centralized political regimes and rigid bureaucracies tend uniformly to subvert the aims of planning. The emphasis here is upon the self-serving tendencies of elites and their capacity to shape public planning through the political structure. In consequence, some have argued that we should return altogether to the private market while others have called for major or revolutionary changes in the political structure itself.

A second reaction has been to call off drastic decentralization so as to create a multiplicity of small communities which will increase the number of choices to individuals and restore an intimate relationship between citizen and political representation. Some of those who advocate this renewal of public choice see in the suburbs a return to small-town America and seek to extend its intimate, democratic forms to our central cities. Thus, opposition to city hall, to the incumbents in Washington and to elite planners finds a responsive chord in diverse American constituencies.

The important contribution of Gottdiener's study of the social management of suburban sprawl is that it provides a strong caution against any simplistic set of alternatives between private and public planning or between centralized and decentralized forms of government. As Gottdiener documents in detail, the suburbs are not so much a series of separate communities which resemble a New England village as they are toy governments lost in a cobweb of larger governmental forms which dwarfs the prestige of their leaders and trivializes local civic participation. The central finding is not the reconstitution of the mythical American community, but the weakness of local political regimes which must scout equally for support among a relatively uninterested local constituency and socially unresponsive builders and real estate people.

The weakness of suburban political regimes not only limits their capacity to plan but also corrupts them and reduces their claim to public responsibility and a broad consideration of public needs. Faced with such weak instruments of policy formation and enforcement, it is understandable that local residents attempt to shape only those decisions that bear most directly on their personal welfare. Thus citizen activism tends to be both episodic and narrow. As Gottdiener points out, the "Privatown" area is a relatively closely integrated submetropolitan region that shows progressive degrees of organizational independence from the central city and integration into a national economy. Despite its economic unity, however, the subregion's governmental forms have remained confused and underdeveloped. Homespun mythology tends to favor the retention of local leaders but economic costs have regularly promoted the farming out of many services to the larger forms of "private government."

It is in this respect that it seems to me that Gottdiener's study is of special importance because it points out that the planning objectives that people seek are primarily those they find within practical reach. It is not that there is a scarcity of idealism in suburbia, or for that matter our inner cities or farm communities, but that there is a grave shortage of public means to implement ideas of broad appeal. Undoubtedly, the bureaucratic forms available fail us and for the most part turn aside those with good intentions. But the more fundamental failing that Gottdiener's study points to are our conceptual shortcomings—our inability to forsee the arrival of submetropolitan regions and to give them some governmental form suitable to the task they face.

Government, of course, cannot be the sole answer, for government must always be informed by some articulate ideology for mobilizing publics behind social policies and programs. The planners' pat notions of economic growth, traffic management, and esthetic order are necessary, but they include no effective or plausible claim that land-use planning will address the essential matters of racial, occupational, or ethnic divisions. So long as the planners avoid such topics, their work is sure to skirt the main fault lines of the society and stay safely away from anything that might capture the imagination of the public or become a mandate for political action. The basic platform of planning, its very fulcrum, is to recognize or anticipate the central strains of a society and to seek some way of accommodating or reducing those strains. As Gottdiener points out, however, the general tendency of planners is to stay largely within the technical vocabulary of physical planning.

The result, as Gottdiener remarks about this submetropolitan region, is that growth has occurred pretty nearly as if there were no planning at all. We have a massive apparatus for planning, but in the end its efforts are largely confined to drawing maps of land-usage after it occurs. But the planners themselves are less at fault than our ideology for guiding societal growth. The leading notion here is that residents are like consumers in the private market, and, if only they can be allowed to have their say, public decisions will be both popular and socially desirable. Yet as Gottdiener shows, many of the suburbs are so balkanized that local residents tend to be extremely defensive even when they are successful in imposing their will. Thus the most frequent consequence is that public participation tends to move the subregion further along toward a pattern of residential segregation and uneven service delivery that have characterized our central cities.

Undoubtedly, citizen participation is important if urban planning is to be an effective and legitimate extension of our public policies. However, it cannot proceed as if it were a mindless organizational process. Over the last two decades, sociologists, planners, and administrators have racked their brains to find a purely structural solution to societal and organizational management. We have repeated over and over the options of decentralization, types of citizen participation, and the advantages of technical competence. In this way we reinvent many of the arguments that have flourished around the topic of reform governments. But if organizations—whatever their character—are to be effective, they must have an articulate and morally defensible set of goals. And they must be endorsed by people of stature whose claims to leadership are grounded in a persuasive moral tradition. Thus, public opinion is not simply the aggregate of individual demands—the model of pollsters and market researchers—but something that must be infused with a conception of the future that transcends parochial and individual interests.

Certainly, there is a shortage of leaders and conceptions of the future which can inform public opinion and enlist it into an effort to plan for a more just and equitable society. But one of the reasons for this scarcity is that we have placed all of our eggs in the basket of organizational tinkering. There is much to be said for such efforts, but alone they leave us without an image of the future that can arouse public opinion and elevate civic participation into something more important than defending one's "propertv values."

These are sweeping generalizations, yet one is pressed to them by this study and its close-grained account of planning in one of our largest submetropolitan regions. The study is richly detailed by case material while also making use of systematic data and documentary sources. As a result the study presents a holistic account of planning which incorporates the complex relationships between citizen participation, political parties, private industry, and popular ideologies. It is the diversity of evidence and approaches which makes the study especially revealing and gives it depth. In this way, Gottdiener makes a strong contribution to an understanding of the emerging shape of our urban community and its public policies.

November 1976
Chicago, Illinois

Gerald D. Suttles

Chapter 1

INTRODUCTION

During the 1950s and 1960s, two images of suburbia were constructed. One, which was critical, emphasized the anonymous sprawl of relatively undifferentiated, homogeneous communities.[1] The second supported a view of suburbia as an untroubled bedroom commuter shed that enabled affluent urban professionals to participate in the new consumer society.[2] By the 1970s, however, we have begun to see that suburbs are neither the ticky-tacky sprawl described by critics nor the untroubled homogeneous havens of the affluent middle class. Certainly, the massive areas around the northeastern metropolitan centers constitute regions and not isolated "suburbias." It makes progressively more sense to call the developed areas adjacent to such central cities "submetropolitan regions."

The societal movements that contribute to metropolitan growth in the post World War II period have come to involve more than community change. Relative affluence and government supported housing programs have created centrifugal patterns of regional development.[3] The suburban areas outside cities have grown on an immense scale. They represent a new form of organizing social activities formerly located in compact urban areas but presently spread out in relatively low-density sprawl. Submetropolitan regions have accumulated a complement of commercial and public institutions that provide a greater degree of independence from the city. At the same time, however, both the central city and the suburbs have grown increasingly dependent upon government and business decision-making which operates at the state and national level. Some of these administrative agents are located in the city, but they are really national or even international in scope.

One of these submetropolitan regions created by the expanding metropolis is Suffolk County in New York State. It extends roughly 30 to 110 miles east of New York City and contains almost 1½ million people. In Suffolk County we can see a growing movement towards cultural and economic independence from the central city and a greater interdependence between it and the other parts of the region. For example, over 60 percent of Suffolk's work force is employed within its borders with an additional 20 percent having jobs in Nassau County, the suburban county directly adjacent to New York City.

With the aging of its housing stock, the social order of Suffolk is moving towards a pattern of internal differentiation which resembles that of our older cities. The region has lost its homogeneity and the trouble-free existence extrapolated for it during the last two decades. Its growing needs include the necessity for the coordination of development and for the supply of social services characteristic of central-city areas. What is apparent now, is that it is also moving in the direction of metropolitan centers with respect to "urban" social problems. Despite ongoing attempts at planning, success has been limited and is not likely to ward off long-term trends towards increasing socioeconomic segregation, pockets of rapid deterioration and suburban blight, the irrational land-use sprawl patterns, the nuisance effects of commercial strip development, gross tax inequalities, government deficit spending, and great differences in public services. Understanding these problems and the process of contemporary development in Suffolk County involves three distinct areas of inquiry. The first calls for a reexamination of our view of the metropolitan region and an abandonment of the narrow view of the suburbs as merely an adjunct to the central city. The second investigates the role of planning and its limited success in coordinating growth within the suburbs. The final area involves an examination of local political organization and the reasons for its particular response to the challenges of a rapidly changing environment.

The Changing Significance of Suburban Regions

Suffolk County was once rural. Over the last twenty years, it has been transformed into a region whose character is best described as "suburban sprawl." The rapid growth is the result of the expansion of the New York metropolitan region and the white exodus from the central city. Seventy-five percent of the county's residents are originally from New York City. Encouraged by federal programs subsidizing housing and transportation, such as the Federal Housing Redevelopment Acts of 1949 and 1954, they left the city in search of more space, greater security, inexpensive housing, and a lower cost of living. Many of these residents first moved to Nassau County, which is roughly one-fourth the size of Suffolk, and which experienced rapid growth in the 1950s. With the exhaustion of the inexpensive housing stock and land in Nassau, the movement east continued into the rural hinterland of Long Island.

The growth of the Nassau-Suffolk submetropolitan region is similar to that experienced by the areas adjacent to most northeastern and midwestern cities. The migration of central city residents to outlying areas is now of greater significance than the shift from the country to the city which was characteristic of the initial stage of urban growth. In fact, a majority of citizens in the United States now live in the suburban sectors of metropolitan regions.[4] One aim of this study is to call attention to this comparatively new form of urban settlement, the submetropolitan region, which is the "emerging city" of the 1970s. The urban planner Hans Blumenfeld has remarked that the shift in the urban demographic center of gravity heralds the first qualitatively new type of human settlement in 5,000 years. Blumenfeld has also called attention to a debate which has arisen over what to call this revolutionary phenomenon. Pop sociologists have called it an urban "explosion," but the real argument is over what is meant by "the city." Traditionally, it referred to a relatively closed, densely populated urban form.

> In speaking of urban revolution we refer today not to the "modern city" but rather to the "modern metropolis." This change of name reflects the fact that from its long, slow evolution the city has emerged into a revolutionary stage. It has undergone a qualitative change, so that it is no longer merely a larger version of the traditional city but a new and different form of human settlement.[5]

A clearer definition of the term "metropolis" can be uncovered through the study of the submetropolitan transformation. Long Island, which comprises Nassau and Suffolk counties, is ideally suited to study this expansion because, as an island bottleneck, it is not subjected to the direct influence of any other urban center. The scale and characteristics of its development patterns represent a relatively pure case of the metropolitan growth of the New York City region. This new form of urban settlement consists of a low-density matrix of businesses, governments, housing and people which extends for almost a hundred miles.

In search of new metaphors for city development a number of urban analysts have included this expansion as part of a *national* network of hierarchical economic, political, and social relationships that have extended the organization of urban social interaction to every geographical area of the United States. This is the "emerging-city" metaphor first coined by Scott Greer. He refers to this process as an increase in societal scale set in motion by innovations in transportation and communications technology.[6] The division of labor in society, the range of communications, and the organizational networks of production and distribution have come to incorporate spatial interdependence on a national scale.

This is certainly the case on Long Island. Banking, commerce, and housing, for example, are dominated by national and international enterprises despite the existence of many local entrepreneurs. Highway construction and the automobile mode of transportation are responsible for the massive scale of metropolitan land-use. In addition to technological innovations, however, the hierarchical scale of the submetropolitan region in Suffolk County is supported by government programs and the utilization of the corporate-bureaucratic form of administration which encourages *mass* development and creates *mass* markets for housing. Our study balances Greer's technological view by investigating the ways in which the social and political postures of local communities in Suffolk County have contributed to the mass mobilization of resources organized on a regional basis which is characteristic of suburban submetropolitan growth.

Our view of the modern city, therefore, retains Blumenfeld's conception of the metropolis, but alters it in two ways. First, we combine his perspective with that of the national urban culture theorists. Organization and control of social activities on a national hierarchical scale utilizing the corporate and bureaucratic form supports regional economies without the necessity of central city dependence. There is, consequently, no sharp organizational boundary to the expansion of the metropolitan region.[7] Despite the recent downturn in home construction, for example, residents in the Eastern towns of Suffolk County, 110 miles from Manhattan, are presently looking for ways to curtail urban growth once construction resumes. They recently sponsored a public referendum which would have enabled the easternmost towns to secede and form a new jurisdictional district, Peconic County. The move, however, was defeated in the New York State legislature.[8]

Second, we wish to take issue not only with Blumenfeld but with most urban analysts on the significance of the suburban submetropolitan region. Throughout the literature on exurban development there is a persistent treatment of the suburban area as an appendage to the central city's social system. Most examinations of growth patterns have used the city as an organizational locus and have considered suburban development as only a consequence of processes which are initiated within the central city. While Suffolk County is affected by activities in New York City, these are more often than not reflections of the latter's function as a national center of business and an impact area of national programs, than of New York City being the focal point of a regional social system. Yet, even the national urban culturist Greer persists in viewing the suburb as a bedroom adjunct of the city. Like many others, he sees the suburbs as the site of the urban middle class, homogeneous, and highly dependent on the central city for its continuity despite the increase in societal scale. The ghost of Wirth and Burgess' concentric-zone theory of urban growth, therefore, continues to haunt the analysis of the suburb.[9]

If this perspective on the suburban submetropolitan region is to be altered, we must abandon the view that it has a homogeneous population and is dependent on the urban center. Following the line of work developed by Dobriner and Schnore, we can see that the Long Island region can be divided into a number of subsectors. Each varies with regard to population density, mix of housing type, independence from the central city, and extent of integration into the national urban culture. There are, thus, many different types of suburban communities, but they are all part of an interdependent region.[10] Suffolk County, for example, contains developments that are heterogeneous with regard to ethnicity, religion, and life style, and contains a population which is relatively independent of central-city employment and cultural amenities. One illustration of regional interdependence is afforded by the new civic arena and entertainment center called the Coliseum. It is the home of both the Islanders hockey team and the New York Nets basketball team. Both had better records in the 1974-1975 and 1975-1976 seasons than their New York City counterparts and have become the center of sports attention for

central-city region		suburban submetropolitan region	
	urban zone of transition		suburban zone of transition
Manhattan	Queens and Brooklyn	Nassau and Western Suffolk	Privatown

← - - - - - - - - - - - - - - - 80 miles - - - - - - - - - - - - - - - →

Figure 1: NEW YORK CITY AND LONG ISLAND SUBREGIONS

many people in the entire metropolitan region. In addition, patterns previously associated with the central city, such as racial and income segregation, are also characteristic of community development in both Nassau and Suffolk counties. Rather than being dependent upon central social processes, Long Island exhibits a *reproduction* of urban-style settlement patterns and social problems tailored to the lower density of suburban land-use and having as its source national trends which encourage large-scale centrifugal development.[11] Figure 1 illustrates the components of this metropolitan region and its present-day geographical scale of growth.

The Limits of Planning

The key problem widely recognized in Suffolk County is the lack of a coordinated pattern of growth despite a heavy social investment in planning.[12] The individual resident confronts a bureaucratic and physical environment which lacks any unity beyond the local activities of everyday life. This is perplexing because there exist professional planning agencies at every stratum of government from the federal level to the unincorporated development. For example, a typical resident of a suburban township in Suffolk County would at the most local level fall within the responsibility of his local township planning board. Yet, since each township belongs to Suffolk County, his would also be included in the county planning board. Both Nassau and Suffolk counties make up the official region known as Long Island, and this region has a joint planning agency called the Nassau-Suffolk Bi-County Planning Board. This was created when it became evident that the individual boards of each county were unable to control the growth of the expanding area. Some of the region is considered part of the New York Standard Consolidated Metropolitan Area which falls within the jurisdiction of the Tri-State Transportation Commission. The New York City region, in turn, is part of the New York State planning agencies, such as the Office of Planning Coordination. These state agencies are included in the planning apparatus of the federal government—the Commerce Department's "northeast corridor" region, and more generally in executive branch planning programs. The hierarchy can be described as follows:

Region	*Planning Agency*
Suburban town	Town planning board
Suffolk County	Suffolk County Planning Board
Long Island	Bi-county planning board
New York City (SCA)	Tri-state transportation commission
New York State	Office of planning coordination Office of regional development
Northeast corridor	U.S. Commerce Department
United States	Executive planning, such as wage-price control boards plus congressional acts, such as HUD 701 assistance to states.

The presence of these multilayered bureaucracies does not appear to impede the process of submetropolitan growth or the reproduction of urban-style social problems. In fact, their presence may make the limited impact of planning on the environment all the more perplexing.

This paradox of the existence of planning agencies and their restricted impact in Suffolk County must be viewed in the context of the national planning experience. Planning in the United States has always been hedged in by a strong commitment to market mechanisms. This state of affairs is most directly attributable to the persistence of the public philosophy of laissez faire. Market forces have been allowed to take over the process of decision-making and have been held to be beneficial to society as a whole if given wide reign.[13] The notion of the limited state is intimately tied with a belief in the sanctified rights of private property.[14] This double-edged resistance to governmental coordination has fostered the faith of localism and the autonomy of the individual's economic decisions which is expressed in the attitudes of both congressional legislators and the judiciary. Municipal efforts at obtaining the power to plan have led to repeated clashes between legal property rights backed up by the courts and the belief in the virtues of unregulated free enterprise supported by urban political constituencies. Despite this, there has probably been more planning in the United States than some of the proponents of the market place would care to admit. The early eastern proprietary cities, such as Baltimore, and capitals, such as Austin, Texas, and Washington, D.C., were all planned by government. The Yankee townships of New England followed the Puritan tradition of active civic pride and participation. Public discussions on aggregate needs were followed by public response such as building roads, hospitals, and schools. The effectiveness of such early civic values provided an ideological model of the virtues of local political control. It led an observer like de Tocqueville to remark that local government in America was founding a democratic society providing for both high levels of participation and individual freedom. He highlighted the tempering of special interests by civic responsibility and organized voluntary participation with broad community goals.[15]

Post-Civil-War America experienced rapid urbanization not unlike present-day submetropolitan expansion. New York City and Chicago, for example, doubled their populations in the post-Civil-War period as a massive wave of European immigration flooded the United States. This gave rise to problems unforseen by the early Yankee town dwellers. By 1868, the social evils of unfettered industrial growth were evident. Trusts controlled the economy, bosses controlled politics, the immigrant work force was exploited and existed at subsistence level, elections were fraudulent, sanitation, sewage, and water facilities were lacking, public education was limited, and utilities were predatory.[16] The de Tocquevillian local community virtues could not manage the scale of urban industrialization. Nevertheless, most individuals still subscribed to religious and moral sentiments regarding civic responsibility and responded by organizing a municipal reform movement.[17]

American city planning grew out of this "turn of the century" voluntary reform movement. In two distinct ways it was subsequently constrained by private-property rights interpreted in favor of individuals by the courts. First, it was reduced to a consideration of physical design. Second, social-reform measures became separated from the purview of professional planners and retained at the discretion of local government. Municipal governments, for example, could not engage in comprehensive social planning which would regulate economic and social activities. Instead, they were confined to the control of city-owned land. They developed this land in a manner to modify the inequities of urban industrialization. Cities annexed undeveloped land for future use and in order to promote or control growth, they developed parks and open spaces to alleviate slum congestion and created civic centers which expressed the aspirations of government in monumental architecture and served as a municipal wedge against private land speculation. After the famous Chicago Exposition of 1893, which featured an integrated spectacle of civic improvements especially built for the occasion by landscape and building architects, these elements became fused in the physicalist "city beautiful" movement.[18] Planning in Suffolk County borrows very heavily from this tradition. Control over land-use and the physical shaping of the growth patterns along landscape principles is the dominant approach of county planning. The head of the Nassau-Suffolk planning board, for example, was trained as a landscape architect.

The social problems of urbanization and the reform measures proposed to alleviate them, such as housing reform, however, have never been widely included in the physical perspective of American city planning. This is as true of Suffolk County today as it was true of early municipal planning. At the beginning of this century the social reformers and voluntary associations became disenchanted with the City Beautiful movement as catering to elite interests. They acknowledged that public works and civic improvements could not alleviate the basic economic and social indigence of the urban poor. Unable to gain acceptance for a broader view of planning from the political apparatus, the reform movement turned to a call for "zoning." This was a land-use control mechanism much more comprehensive in scope than landscaping principles and modeled after the rational ordering of societal activities found in the city ordinances of the German towns at the time. The technique was called "comprehensive planning" as it enabled the municipality to exert some control over private market and public decisions and to regulate housing, transportation, and the location of industry in an overall framework coordinating growth.[19]

In 1909 a national conference on city planning was called and it was felt at the time that a truly comprehensive social planning ethic, not unlike Fabian Socialism, would be forged in this country and that it would temper free enterprise and private property rights in order to promote the general public welfare. Interest in zoning was greatest in New York City which had a strong constituency led by Henry Marsh, whose group had organized the conference, and it was there that the first citywide municipal zoning ordinance was adopted in 1916.

Despite this reform movement and its early success, the federal government did not take the initiative in developing a comprehensive social planning function utilizing the zoning mechanism. It reinforced, instead, the traditional sanctity of private property rights. Decisions on land-use policy and zoning were left to the judiciary, which only reluctantly heard a few cases, and land-use is still one of the least developed areas of litigation. While there are several exceptions, federal court decisions have usually left land-use policy to the community. In turn, state courts have usually joined the federal ones in trying to avoid zoning cases.[20] It was not until 1926, after almost a century of urban industrial growth, that the Supreme Court upheld the constitutional validity of zoning by municipalities in the case of "Euclid vs. Amber Realty." Two years later, however, in the case of "Nectow," the Supreme Court established the right of an individual landowner to obtain judicial relief from "particular applications of that valid principle."[21] The measures to be taken by citizens in obtaining that relief have never been explicitly spelled out. No clear initiative was taken for comprehensive planning by *any* branch of the federal government, and the autonomy of home rule was reinforced. The power of land-use control was effectively separated from the planning function of national government.

The consequences of this separation can be seen more clearly in the developments in planning after World War II which are most directly relevant to Suffolk County. The main efforts have been through federal programs in housing and transportation and through the creation of a plethora of planning agencies at almost every governmental level. These two developments are somewhat related. Following World War II, the United States, like the European countries, suffered from a severe housing shortage created by the priorities of war mobilization. The Congress passed legislation which actively subsidized the construction of new housing and highways. The Federal Housing Redevelopment Acts of 1949 and 1954 guaranteed mortgages and subsidized veterans with loans at low interest in addition to supporting urban renewal. The Federal Aid Highway Act of 1956 established the 40,000 mile "National System of Interstate and Defense Highways" to be supported 90 percent by the federal government and 10 percent by the states through a national excise tax on gasoline. Together these acts opened up the urban hinterland to development and created a mass market for new housing. They set in motion the creation of submetropolitan regions developed on the scale of Nassau and Suffolk counties. In 1949 and every year of the 1950s, more than a million units of housing were built.[22] This was a comparatively successful record relative to the accomplishments of the other industrial countries. During the same period, for example, in England under a nationalized policy of land-use and social planning, only 300,000 units were built each year despite a larger demand.[23]

The post-war legislation, however, was notable for its lack of initiative in formulating programs to help coordinate growth. Its Keynesian market supports left decision-making to local areas and the private marketplace without the benefit of any compelling regional framework or integrated programs designed to guide local municipal decision-making. Ironically, the uncoordinated sprawl pattern of development occurred with the full benefits of foresight. As early as 1948, the congressional hearings on the housing bill were informed by professional planners of the need for a national land-use policy to accompany the housing legislation. In 1950 the American Institute of Planners warned against uncoordinated metropolitan expansion and proposed alternatives modeled after the British New Town legislation.[24] In response, Congress appeased the planning profession by requiring each local area requesting aid for housing and urban renewal to produce and submit a comprehensive "master plan" for development. Although failing to initiate legislation which would have direct powers of implementation, the requirements for federal aid supported the establishment of planning agencies at every municipal level and created jobs for the profession of city planning.[25]

It is the existence of these organizational levels of land-use planning which present a paradox. For, while the sheer quantity of housing constructed in the United States is impressive, its limited availability to the lower-income groups, the sprawl of faceless suburbs, and the beginning of patterns of segregation and injustice characteristic of central cities found in these new submetropolitan regions, raise serious doubts about the effectiveness of government planning. In fact, much of the legislation and local planning may have been counterproductive because it often provided the tool by which segregation was insured or land-use objectives constantly eroded by legal and administrative clauses.

The failure of governmental initiative to implement broad policies of coordination in the public interest has been matched at the local level. Such weak response leaves the pursuit of social goals in the hands of reformers and channels the redress of private sector inequities into the hands of a reluctant judiciary. One consequence of this has been the generation of an incredible variety of often inconsistent rulings on local land-use so that no aggregate regional policy is in evidence. There seems to be no capacity to shape local growth to avoid the accompanying problems of submetropolitan expansion. A second consequence is that local home rule has at times utilized the zoning mechanism as a means to support the local political organization divorced from its planning function. In Suffolk County there is a clear separation between land-use planning which is carried out by professionals or local holders of patronage in an advisory capacity, and local political control of zoning. The inconsistencies, which exist nationally, and the use of zoning as a defensive and segregating measure are as evident in Suffolk County as is the lack of initiative on the part of local government to utilize its land-use controls for social-planning purposes.

The Weakness of the Local Political Organization

Our study of Suffolk County replicates the national experience in many ways. In subsequent chapters we shall fill in the details of this process. As a study of the New York metropolitan region there are, however, a number of distinct aspects which differ from housing patterns found elsewhere. On Long Island there is little individual construction of housing. In other areas many people build their own homes utilizing local resources or using the labor of family and friends, but in Suffolk County restrictive building codes in conjunction with zoning ordinances make this especially difficult. This clears the way for large organized businesses to dominate home construction. In several areas of the country individuals can still deal with each other securing permits, enlisting the aid of acquaintances, and securing the services of small businessmen. On Long Island we have comparatively large organizations dealing with each other. Corporate home builders interact with unionized labor and the local political parties. The legal and governmental constraints on house construction in the region, therefore, lend themselves to the bureaucratization of the home construction industry and have resulted in a situation which greatly restricts the operation of market forces or free entry.

In addition, the nearness to New York City's mature banking and corporate institutions and the scale of the urban exodus encouraged by national programs has resulted in the creation of a mass market for housing. Most homes are built by a small number of large firms and are financed by a select group of banks, whose operations fuel the rapid growth of the area. If the profits to be gained through the harnessing of this mass demand are considerable, the losses through excessive delays in obtaining the necessary permits from local government are equally immense. Lacking other resources, the leaders of political organizations have learned to utilize the bureaucratic constraints on home construction for party support, and, as a consequence, building and land-use in the Long Island submetropolitan region is politicized as well as bureaucratized. Developers and speculators must bargain with local politicians who act not always in the public interest. Our study of Suffolk County sheds light on the relationship between government planning powers and their political use. It reveals how control of public decision-making can be used as a bargaining device for negotiations with the limited number of corporations in the private sector for party and personal gain. In effect, planning agencies, the large scale housing industry organizations and the weakness of local party support lend themselves especially to a political process that bargains away land-use policies for party loyalties.

Our study of the limits of planning suggests that ineffectiveness is due to the weakness of local political parties and regimes. Whatever the inadequacies of national planning legislation, these have been matched by the failure of local government to rise to the challenge of harnessing growth in the public interest. In some middle-sized cities, such as Minneapolis, municipal coordination of growth has not been coopted by narrow special interests.[26] In Suffolk County, however, there is an absence of any articulated process which allows for the aggregation and expression of shared public opinion that takes a broader view towards growth. Planning is not utilized to meet the defensible social aims of the region, but is more often than not a reflection of special interests which are seldom linked to the needs of other areas. Heavily fragmented by jurisdictional lines, lacking a stable constituency, and not in possession of much patronage, both political parties and elective officials have been largely dependent upon the real estate industry as their persistent source of support. This has of necessity made suburban politicians and administrators poor planners and ineffective agents in enforcing the goals of planning or conceptualizing the implications of the scale of submetropolitan growth.

Various factors contribute to the weakness of local "toy" governments and party politics. First, the suburban polity is issue-oriented and lacks the strong party commitment which is characteristic of central city machines. The labor unions, municipal workers and long-term residents which make up many of the party supporters in cities do not exist actively in suburban townships. Submetropolitan area residents engage in much ticket-splitting. For example, Suffolk County is widely known as a Republican stronghold, and yet voters regularly return Representative Otis Pike, a Democrat, to Congress. Political apathy is also expressed among voters except in brief periods of reform. There is a significant and probably growing proportion of independent voters and people who simply fail to register. Support for local parties is also weakened by the quagmire of fragmented districting in the region.[27]

Second, the "limited liability" of many new residents is matched by the limited capacity and ambition of local political leaders. Politicians with their "toy" governments cannot do much. The relative newness of the population as a result of rapid growth has forced old political sentiments and relational ties into the background of township politics. The parties presently suffer from an inability to attract prominent citizens to run for office despite the presence of such individuals. In Suffolk County politicians tend to be real estate or insurance agents, lawyers or local businessmen. The party is organized by small units almost at the street level, and politicians address needs at that level, striking up limited, almost contractual, bargains without guidance from underlying values or an integrating political ethic. In fact, in Suffolk County there is no clear conception of what the function of local government should be beyond the vague notion of "fulfilling the public trust," and, we may add, the protection of property values.[28]

Third, the restricted view of the role of local government is reinforced by the low visibility local politicians have in state and national politics. News media tend to spotlight central city politics, some fifty miles away, while suburban areas are rarely given the same treatment. The lack of news coverage given to this submetropolitan region restricts the ambitions of local politicians for state or national offices. Long Island, at least, has its own daily newspaper, *Newsday,* so that it is better off than many suburban areas. *Newsday,* has an excellent reputation and does provide broad coverage to Nassau and Suffolk residents of the full scope of daily events. TV, however, is city based, and along with the city newspapers, such as the New York *Times,* it often ignores the suburbs except for the more ludicrous examples of township politics. Consequently politicians are viewed by the suburban polity more as local party hacks operating in self-interest than as professionals pursuing a public career. Periodic exposures of corruption reinforce this negative view and further help to restrict patronage and the ability to recruit widely among political aspirants. This prepares the way for future corrupt practices in order to support local government as the political resource base shrinks.[29]

Without an effective polity, the best laid plans and the most carefully worded planning legislation are of no avail. Undoubtedly, the need for federal land-use policy and enforcement is great. In addition, however, our study of local government suggests that there needs to be a recognition of the submetropolitan region as an emergent social reality and some effort to give that region the community identity and common fate that will make planning a desirable and achievable goal. This requires a clear articulation of the functions of local government. At the present there is no aggregating connection between unincorporated areas which contain the bulk of the suburban communities, along with the villages and the townships which comprise the county, so that the expression of interests and sentiments do not move beyond the immediate neighborhood level. This is well below the critical stage of shared interests necessary for the liberal politics of compromise and pluralist bargaining to function. At present, the incapacity of local government in Suffolk County is so great that it cannot implement the broad conceptions of social welfare that are necessary to viable planning and the coordination of growth.

This study addresses the problems of planning and the limitations of local government in a submetropolitan region. The extent to which the experience of Suffolk is shared by other areas is uneven, but it is reasonable to suggest that many large U.S. cities are undergoing the same process of development. Such regions no longer lead an untroubled existence. They carry the burden of coordinating the increased scale of growth encouraged by national programs. They face a social resistance to an interventionist government, especially one which restricts the use of private property. Yet, the inequities of the market place and the minimally regulated activities of the real estate interests compound the problems of coordinating development. Finally, the vast regions opened up to metropolitan expansion lack a visible political structure for aggregating local citizen needs and articulating shared sentiments that would be broadly based and in the public interest. Citizen participation in the political process is ad hoc and fragmented. The looseness of response and the lack of any underlying vision of growth is matched as well by the local politicians and businessmen. The result is that the suburban environment has taken on a shape which is becoming increasingly difficult to understand and plan.

We will examine in some detail the characteristics of this development pattern in Suffolk County. In addition, we will focus on a single township, Privatown, as the site of our case study. It is one of the ten townships comprising the county. Its area is roughly that of Nassau County and it is the largest township of Suffolk. Furthermore, Privatown contains the suburban fringe area and was the subsector which experienced the most rapid growth during the period between 1960 and 1970. We will detail the process by which local government; national and local businesses; regional, county, township, and advocacy planners; real estate speculators; and local citizen groups interact as development proceeds, in such a fashion as to bargain away the planning objectives that were mandated and intended to serve broad conceptions of social welfare. In addition, we hope to show why government has become larger but not necessarily more effective, and, as a corollary, why more planning in the United States may not necessarily meet the needs of the present majority of people in coordinating the activities of their submetropolitan environment.

NOTES

1. William H. Whyte, Jr., *The Organization Man,* Garden City, N.Y.: Doubleday Anchor, 1956; John R. Seeley, R. Alexander Sim, B. W. Looley, *Crestwood Heights,* New York: Basic Books, 1956, A. C. Spectorsky, *The Exurbanites,* Philadelphia: Lippincott, 1955.

2. George Lundberg, et al., *Leisure, A Suburban Study,* New York: Columbia Univ. Press, 1934; Sylvia F. Fava, "Suburbanism as a Way of Life," *American Sociological Review,* 21 (February 1956), pp. 34-37, Scott Donaldson, *The Suburban Myth,* New York: Columbia Univ. Press, 1969, Robert C. Wood, *Suburbia–Its People and Their Politics,* Boston: Houghton Mifflin, 1959.

3. Basil G. Zimmer, "The Urban Centrifugal Drift," in Amos Hawley et al., *Metropolitan America in Contemporary Perspective,* New York: Halsted Press (A Sage Publications book), 1975. See also the introduction to this volume, for a summary of trends.

4. According to the 1970 census, suburbanites constituted 37.6 percent, and central city residents were 31.4 percent of the U.S. population.

5. Paul D. Spreiregen (ed.), *The Modern Metropolis: Selected Essays by Hans Blumenfeld,* Cambridge, Mass.: MIT Press, 1967, p. 61. Blumenfeld rejects the term "conurbation," used by Geddes and Mumford, because it implies formation by the fusion of several pre-existing cities, and the term "megalopolis," coined by Jean Gottmann, because it refers to the entire region of the Northeastern seaboard. Other technical terms have been suggested. For example, Jerome Pickard's "Urban Region" and Brian J.L. Berry's "Daily Urban System," cf., U.S. Commission on Population Growth on the American Future, *Population, Distribution and Policy,* Washington, D.C.: Government Printing Office, 1972, p. 142, 240. Descriptive terms have also been suggested for the new urban form, such as spread city, polycentric city, radial city, etc.

6. Scott Greer, *The Emerging City: Myth and Reality,* New York: Free Press, 1962, Scott Greer, *Governing the Metropolis,* New York: John Wiley, 1962. The term "increase in societal scale," was coined by Godfrey Wilson and Monica Wilson in *The Analysis of Social Change,* London: Cambridge Univ. Press, 1945. Although Greer indicates the organizational aspects of this process by discussing the role of corporate and bureaucratic forms, his argument relies heavily upon technological innovations for causality and, therefore approaches a form of technological determinism. In processes of social change, social organization and technology are always

interrelated. For an excellent demonstration of this approach, cf., Raymond Williams, *Television: Technology and Cultural Form,* New York: Schocken, 1975.

7. This massive pattern of geographical growth is referred to as "megalopolis," cf., Jean Gottmann, *Megalopolis: The Urbanization of the Northeastern Seaboard of the United States,* New York: Twentieth Century Fund, 1961.

8. November 11, 1973.

9. For examples of contemporary recasting of these same models, cf., Brian J. Berry, *The Human Consequences of Urbanization,* New York: St. Martin's Press, 1973, and Claude Fischer, "Urbanism as a Way of Life–A Review and An Agenda," *Sociological Methods and Research,* (November 1972), Vol. 1, No. 2, pp. 187-242. In an empirical test of the Burgess model, Schnore and Klaff found that the older metropolitan regions tended to conform to it, while the newer areas of the Southwest did not, cf., Leo Schnore and Vivian Klaff, "The Applicability of the Burgess Zonal Hypothesis to 75 cities in the U.S.," Madison: Center for Demography and Ecology, University of Wisconsin, Working Paper 72-78 (November 1972). In many European cities, such as Paris, the suburbs are the site for workers' housing, while the central city contains the more affluent members of the population.

10. William Dobriner (ed.), *The Suburban Community,* New York: Columbia Univ. Press, 1969; Leo F. Schnore, *Class and Race in Cities and Suburbs,* Chicago: Markham Press, 1972. See also Charles Haar (ed.), *The End of Innocence: A Suburban Reader,* Glenview, Ill.: Scott, Foresman, 1972, and David Thorns, *Suburbia,* London: MacGibbon & Kee, 1972, for a general discussion of this kind, and the case studies: Bennet Berger, *Working Class Suburb,* Berkeley: Univ. of California Press, 1960; Herbert J. Gans, *The Levittowners,* New York: Pantheon, 1967; and William Dobriner, *Class in Suburbia,* Englewood Cliffs, N.J.: Prentice-Hall, 1963.

11. For further discussion of these general trends, cf., Haar, Thorns, ibid., and Louis Masotti and Jeffrey K. Hadden (eds.), *The Urbanization of the Suburbs,* Beverly Hills: Sage, 1973; and *Suburbia in Transition,* Chicago: Quadrangle, 1974.

12. Most observers of growth in the United States see it as chaotic and unplanned. There is, in fact, much social planning in the United States. Consider the programs and subsidies created by federal housing legislation: *The U.S. Housing Act of 1937; The Federal Housing and Redevelopment Acts of 1949, 1954, 1968, 1970,* Title I, Urban Renewal Support from the 1949 act; The Federal Housing Administration; Section 221.d.3., 1961; Sections 235 and 236 and Title IV of the 1968 act; Title VII of the 1970 act. In addition, there has been employment legislation, such as the *1946 Employment Act,* and highway construction programs, such as the *Federal Aid Highways Act, 1956.* We shall address the limitations of the present system of planning more fully in the following chapters.

13. Even socialist societies recognize the value of market processes as decision-making mechanisms. Oskar Lange uses the marginal pricing system in his socialist scheme, while the Soviet economist, Lieberman, uses a mathematical version of profits to measure productive efficiency, cf., Oskar Lange, *On the Economic Theory of Socialism,* Minneapolis: Univ. of Minnesota Press, 1938.

14. See, C. B. MacPherson, *The Political Theory of Possessive Individualism,* New York: Oxford Univ. Press, 1962.

15. Alexis de Tocqueville, *Democracy in America,* New York: Schocken, 1961.

16. For a comprehensive history of planning in the United States, see, Mel Scott, *American City Planning,* Berkeley: Univ. of California Press, 1971.

17. Among those most famous for activities in the settlement house movement were Jane Addams in Chicago, and Lilian Wald and Jacob Riis in New York City.

18. Scott, op. cit.

19. Cf., J. Delafons, *Land Use Controls in the United States,* Cambridge, Mass.: MIT Press, 1969.

20. For discussions of the court's role in planning, cf., Charles Haar, *Land Use Planning,* Boston: Little, Brown, 1959; Richard Babcock and Fred Bosselman, *Exclusionary Zoning,* New York: Praeger, 1973; Bernard H. Siegan, *Land Use Without Zoning,* Lexington, Mass.: D. C. Heath, 1972.

21. Babcock, op. cit.

22. Marion Clawson and Peter Hall, *Planning and Urban Growth,* Baltimore: Johns Hopkins, 1973.

23. Ibid., the authors compare the postwar British and U.S. experience. See also Marion Clawson, *Suburban Land Conversion in the U.S.,* Baltimore: Johns Hopkins, 1971.

24. Scott, op. cit., p. 474.

25. Between 1958 and 1971, for example, the total enrollment in U.S. planning programs went from 586 to 5,119. In 1971 there were 83 separate programs in planning in U.S. schools and an additional 71 in related fields, cf., "ASPO's 13th Annual School Survey," American Society of Planning Officials, 1972. In 1969 The American Institute of Planners listed 8,000 individuals as planners in the United States. This is a very conservative estimate of all those involved in the profession, because many individuals work in planning bureaucracies but do not have the professional title of planner.

26. For examples of the function of planning within a pluralistic context, see, Alan Altschuler, *The City Planning Process,* Ithaca, N.Y.: Cornell Univ. Press, 1965; and Robert A. Dahl, *Who Governs?,* New Haven: Yale Univ. Press, 1961.

27. See Chapter 4.

28. Ibid.

29. Ibid.

Chapter 2

THE SUBMETROPOLITAN TRANSFORMATION

New York City metropolitan expansion into outlying areas was first a process of suburban community development. The housing stock was new, and it was built and marketed on a mass scale. This centrifugal pattern of growth was facilitated, since the 1950s, by the construction of an extensive highway transportation network. This includes both the northern and southern state parkways, which are limited access roads that run along either shore, as well as the Long Island Expressway, a commercial six-lane highway, which extends across the central industrial spine of the island. Together these roads form a grid with two- and four-lane north-south intersecting highways, such as the Meadowbrook Parkway. This net allowed home construction to progress rapidly eastward from New York City, first in Nassau, and then in Suffolk County as farmland became converted to the familiar sprawl of tract housing. Such development took place on an immense scale, extending for at least sixty miles from Manhattan.

The early stage of this transformation in land-use gave grounds for the notion of suburbia as a series of isolated, upper middle-class "bedroom communities" characterized by cultural and socioeconomic homogeneity.

The new developments existing in unincorporated areas were fused with the older, exurban incorporated villages into a "suburban ring" of the city. Initially, the suburban communities were rightly considered a single part of the urban landscape much like the ethnically homogeneous "natural areas" of the central city or its economically exclusive enclaves, such as its "Gold Coast." Consequently, suburbia became the preoccupation of an urban analysis which relied very heavily on the city-suburb contrast.[1] At the time, it seemed clear that the suburban ring around the central city offered an identifiable "suburban way of life" characterized by an emphasis on the nuclear family and middle-class child-rearing patterns, a preference for lower-density urban life, and an outspoken passion for local government of restricted size and scope.[2]

The lack of old housing stock and commercial establishments was, however, temporary. At present, this situation has altered drastically, and it is likely to change more in the near future. New businesses have arisen that cater to the needs of the greatly increased population, which has become progressively more heterogeneous in composition. Differential aging of the housing stock and differential success at local community improvement have resulted in an increasingly stratified set of social areas. Economic and social differences in Suffolk County are becoming more evident as are the existence of poverty pockets and suburban ghettos, as regional development has had some time to mature. Several communities have also been bypassed entirely by the prosperity of the past decade, and are now the scenes of suburban blight.

This growing maturity was already acknowledged in the 1960s by the suburban studies which attempted to alter earlier assumptions. The key points of attack focused on the propositions of regional homogeneity and the usefulness of the city-suburb dichotomy as an analytical device. Long Island communities contributed as objects for part of this research endeavor.[3] The city-suburb contrast fell reluctantly before a new urban analysis which recognized the increasing complexity and interdependence of the submetropolitan region. Suburban community diversity, variations in industrial development, income, occupational and cultural characteristics were widely recognized.[4] At the same time, researchers developed the notion of a metropolitan process of growth in which development increasingly pointed toward what was referred to as a "super-city" or "megalopolis."[5]

In 1971, the Nassau-Suffolk submetropolitan region was declared a separate Standard Metropolitan Statistical Area (SMSA) by the census bureau, giving federal recognition to the new low-density form of urban settlement.[6] This bi-county Long Island region adjacent to New York City has almost three million residents. It is comprised of 897 units of government, including two counties, two cities, 13 towns, 92 villages, 135 school districts, 126 fire districts, 477 town special districts, 16 public authorities, and 11 urban-renewal districts.[7] It is reasonable to expect that submetropolitan regions adjacent to other U.S. cities will also grow to this size and complexity and become future candidates for SMSA status. In what follows, we shall bring into focus this change in perspective and argue for a regional view which assumes growing complexity, socioeconomic diversity, and the increasing interdependence within the submetropolitan social system.

Relative to New York City, the Long Island region is more affluent and contains a greater proportion of white people. However, the various subareas within this extensive development, such as Suffolk County, and its townships, such as Privatown, must be studied on their own terms. The growth patterns of this submetropolitan region suggest that we must explore issues relating to the extended process of suburban growth from the 1950s to the present.

The Growing Complexity of the Submetropolitan Region

Presuburban Suffolk County was rural. Land was used mostly for agricultural purposes and heavily dominated by New York City's economy. In total product cash value Suffolk is still first in New York State and among the top ten countries in the nation. Long Island potatoes and ducks are but two of the important crops. In presuburban days, the county was also the site of large estates belonging to the wealthy urban dwellers. The mansions of millionaire industrialists dotted the north and south shores. Southampton attained its leisure-class cultural peak in the pre-World War II days of F. Scott Fitzgerald. In addition to agriculture and recreation, Suffolk contained a number of government defense-related industries, for example the World War I camp Upton, which later became the Brookhaven National laboratories, and private defense contractors such as Grumman and Republic. Over the last two decades, the county has shifted from being an agricultural-recreational-rural area to one that is dominated by white-collar employment, retailing-service-oriented industry, education, and diverse styles of life. The process of this submetropolitan change can be shown best by examining the new demographic, housing stock, and industrial patterns which altered the area between 1950 and 1970. Our discussion of this transformation will also demonstrate the significance of both Suffolk County's growing independence from the New York City economy and its new regional and national interdependence.

The key factor of growth has been the migration of the central city population. The massive movement of people set in motion an ecological adjustment process which is still incomplete. In 1970 Suffolk County had a population of 1,127,030, an increase of almost 70 percent over the 1960 census. Suffolk's neighbor Nassau, which is nearer the city, had a 1970 population of 1,428,838, but this reflected only a 10 percent increase over the 1960 level. While 74 percent of Suffolk's population increase was from migration, only 15 percent of Nassau's increase could be accounted for in this way. Thus, between 1960 and 1970 the transformation from a rural to a residential community was fast under way. The fringe area of this transformation is presently located in the township of Privatown, whose population more than doubled in the same period.[8]

This phenomenal rate of migration corresponds to the demographic shift occurring simultaneously in New York City. Between 1950 and 1970 the *white* population of the city declined by almost 2 million. Approximately 33 percent of these left New York City for Suffolk County.[9] The reasons for the out-migration of whites include: the availability of relatively inexpensive housing, which could be purchased under "GI Bills," the Housing Acts of 1949 and 1954; FHA guidelines and HUD 235 subsidies; the relative affluence enjoyed by middle American blue- and white-collar workers due to increased defense spending of the Viet Nam years; racism and the block-busting techniques of central city speculators; the urban middle- and working-class dislike for the city and their pursuit of a more privatized family life, including the single-family home on Long Island. Contrary to earlier observations, which centered upon central city decay, recent attitudinal surveys of Suffolk County residents reveal that the decision to move is shaped more by "the pull of the suburban region than the push of the city."[10] A 1973 study indicated that former city-dwellers came to Suffolk primarily in search of a more attractive environment to raise a family and, second, because of the relatively low cost of homes in the area.[11]

In this development pattern of regional growth the suburban fringe which represents the interface of agricultural to residential land-use conversion is always the site of the least expensive dwellings. The filter effect of home purchasing and construction operates throughout the metropolitan land mass, affecting both the central city and the suburban subsectors closer to it as well as pushing the fringe area further east on the island.

Thus, Suffolk County developed, as residential land became relatively unavailable and high priced in Nassau County, and as the demand for housing was sustained. This growing movement eastward of people and homes took place in a multistaged process that included, in many cases, a move from the city to Nassau County first. This filter pattern is often called "trading up," because residents cash in on the equity acquired through home ownership in rapidly growing areas by purchasing less expensive new housing at the fringe with the inflated sale price of their old one located closer to the city. A study done on a sample of Privatown homeowners in 1968 revealed that

> only 11% of the respondents have lived in their present home since marriage. A plurality of 38% of the women have had 2 homes since marriage, 22% had 3 different homes, and over 25% have changed addresses four or more times–including six women (3%) with seven or more residences.[12]

The multistepped nature of such mobility was also revealed by the fact that only 18 percent of the respondents were born in Privatown. The remainder of the sample included 50 percent former New York City residents and 15 percent who were born in Nassau County or other parts of Suffolk.[13]

Between 1950 and 1970 inexpensive housing was available in Suffolk County. Due to government subsidies, such as FHA programs, the ease of getting financing, and tax incentives it *paid* to own your own home during this period. In addition, due to the rapid nature of growth it appears that incentives were also created for the massive "trading-up" filter effect of home ownership. One implication of the studies on resident mobility concerns a possible source for the limited attachment to their community observed in suburban homeowners. The same 1968 Privatown survey, for example, revealed that most residents had little identification with their township, county, or community. When asked where they lived, 50 percent of the respondents in the survey answered Long Island, only 27 percent said their own community, 15 percent said Suffolk County and 2 percent said Privatown.[14] This "limited liability" of mobile residents has important consequences for the support of local township politics.

Presently, it is becoming increasingly difficult for all but the relatively affluent to afford new housing in Suffolk County. In part, this is due to inflation, which has raised the interest rate greatly, to soaring property taxes, and to the increased costs of home construction. It is also due to energy constraints, which have raised the price of home and automobile fuel. There is no question that the development of the suburban submetropolitan region is reaching a new phase of growth which may differ from the initial expansion period. Due to the higher costs of all suburban housing and the high cost of living required to sustain life at the fringe, we may be witnessing a decrease in the rate of metropolitan expansion. This slowdown, however, is taking place at roughly *eighty miles* from the center of New York City.[15]

The demographic shift to Suffolk County altered the stock of housing. In fact, changing categories of land-use are the clearest indicators of the submetropolitan transformation of a rural region. The share of owner-occupied and multifamily housing increased markedly, while the number of renter and single-family housing units remained roughly the same (see Table 2.4). Figures for owner occupancy indicate that more families live permanently in the area. In Suffolk 74 percent of all housing is owner occupied. This reflects the shift from a second-home recreational area to a first-home residential one which is characteristic of Suffolk County. Presently, in the bi-county region as a whole, 82 percent of all residents own their own home, which compares to 63 percent for the nation, and 84 percent of all housing units are for single families, which compares to 74 percent nationally.[16]

The increase in the percentage of multifamily homes is another indicator of changing suburbanization. As more families move into an area, the need for new housing outstrips the supply. Some residents move into previously vacant housing, but, in response to the demand for relatively less expensive shelter, apartments are also built. Uncertainty in employment, the growing number of young couples, and the demand for aged and low income workers contribute to an increasing need for apartment units. A mature suburb comprises a surprisingly large proportion of multifamily dwellings. Eleven percent of Suffolk's stock is devoted to this type of housing, whereas the more mature housing stock of Nassau County includes over 20 percent multifamily dwellings.[17] Furthermore, it appears that in a suburbanizing region the demand for new housing prompts home owners to remodel underutilized space in their homes for use as illegal apartments not reported to county or tax authorities. Suburbia, therefore, contains a diverse mixture of housing and a need for apartments, which official statistics do not totally reflect.

The need for housing, fueled by rapid growth, is also reflected in vacancy rates. Table 2.1 shows a decrease of 24 percent in the number of vacant homes. Such housing absorbs a significant number of recent

Table 2.1: Changes in the Stock of Housing for Suffolk County, 1960-1970

	1960		1970		1960-1970	
Housing Type	Total	% of Total Housing	Total	% of Total Housing	Change	% Increase
Total housing units	224,451	100%	335,041	100%	+11,590	+45%
Owner-occupied	146,946	65	245,818	73.4	+98,872	+67
Renter-occupied	26,466	12	49,769	15	+23,303	+90
Vacant housing	51,039	23	38,789	11.6	−12,250	−24
Single-family	178,573	78	276,575	80	+98,002	+55
Multi-family	12,021	5.4	33,396	10.8	+21,375	+178

NOTE: U.S. Census of Housing, 1960 and 1970.

arrivals, who rent or buy previously unoccupied dwellings. In 1970, the combined bi-county vacancy rate was 1.4 percent which is considered "extremely tight by federal standards."[18] The recent slump in new housing starts, the widespread use of exclusionary zoning, the restrictions on multifamily apartment construction, which lowers density of land-use, and the high demand for all types of housing in the region have also raised the costs of owner- and renter-occupied units. Recently, these factors have contributed to a growing shortage of moderate-priced housing in the area, which is expected to become more acute in the future. In 1970, for example, the dominant range for new homes was $35,000 to $50,000 in Nassau and western Suffolk, while the dominant rental range for one-bedroom apartments on Long Island was $215 to $260 a month.[19] At present, only the relatively affluent can afford such prices. On the one hand, this has caused a severe problem for moderate- and low-income residents already living in the region, and, on the other, constrained the ability of moderate- and low-income residents to move to the suburban region.

In Suffolk County the change in housing stock and in land-use, the disappearance of estates and exclusive recreational areas, and the growing housing shortage reflect the ongoing process of submetropolitan transformation. They illustrate the way in which residential developments, extending for miles, replace open farm land, dotted by small "exurban" villages, as a rural-recreational region becomes a suburban one.

The Suburban Economy

Large scale migration rapidly altered the function of the regional economy. In addition to changes in the composition of the work force, Suffolk County benefited from the centrifugal growth of New York industry and commerce as jobs followed people in the affluent 1960s. Each new family that arrived represented an increase in the demand for goods and services. Retailing and commercial shopping centers arose to harness the market of these voracious, family-oriented consumers. These centers, in turn, created new jobs for housewives and young adults, which contributed to the spending multiplier effect and added to family disposable income. Over time, the independence of the regional economy from the city emerged, as reflected in journey-to-work statistics. Ties to national and international corporations and banks replaced the limited rural economy with its locally controlled businesses.

CHANGE IN LABOR FORCE AND CONCENTRATION OF INDUSTRY

In Table 2.2 we see that over a twenty-year period blue-collar employment in the county declined, agricultural work almost disappeared, and white-collar and service jobs increased in importance. The suburban transformation created a majority of white-collar workers in Suffolk County, and on Long Island three out of five workers presently have white-collar jobs. Employment in the service sector was generated as a concomitant of this demographic shift, and accommodated the high-consumption wage earners.

Work-force changes are also reflected in shifts in industrial importance. This is revealed in Table 2.3, whhich illustrates the sharp decline in Suffolk County agriculture.

Table 2.2: Breakdown of Labor Force by Industry

	Percentage of Labor Force		
Suffolk County Industry	**1950**	**1960**	**1970**
White-collar	39	44	51
Blue-collar	40	39	32
Service	31	11	16
Farm	4	1	0.5

NOTE: U.S. Census of Population, 1950, 1960 and 1970.

Table 2.3: Census of Agriculture for Suffolk County

	1950	1959	1964	1969
Acres	123,346	89,776	74,808	61,520
Number of Farms	2,187	1,258	1,138	743

NOTE: U.S. Census of Agriculture, 1970.

Nationally there has also been a shift in industrial importance to service categories. Table 2.4 illustrates this for the metropolitan area and shows that the growth of service payrolls is greatest in the suburban areas. The entire region experienced an increase in service industries, but compared to New York City, Suffolk received a disproportionate amount as its share. Second in magnitude of increase was the retailing sector of the economy. Here too, Suffolk's increase was disproportionately large when compared with the entire metropolitan region.

The economy of the suburban submetropolitan region is characterized by the predominance of the retailing, service, and light industrial sectors. These features have been accentuated recently by an absolute decline in blue-collar employment. Defense contracts, the primary source of such jobs, diminished in the 1970s. In fact, unemployment is at a considerable level in Suffolk County; as of September 1975 it was 8.3 percent in the bi-county region, 1.5 percent above the national average. With recent rises in the cost of living, this has contributed greatly to the growing differences in socioeconomic status. In addition, the recent recession has hit the region especially hard and has contributed to the slowing down of its growth rate.

Table 2.4: Percentage Change in Payroll Employment in New York SMSA and Suburbs, 1959-1965

	Total	Total Suburban		
Industry	NYSMSA	SMSA	Nassau	Suffolk
Manufacturing	1	15	27	15
Retailing	11	40	40	89
Wholesaling	4	66	99	62
Construction	4	24	31	60
Finance, insurance	7	51	78	55
Services	26	58	73	143
Total for all industries	9%	37%	42%	55%

NOTE: Residential Market Analysis, Part II, Nassau-Suffolk Bi-County Planning Board, 1968, p. 6.

THE INDEPENDENCE OF EMPLOYMENT FROM NEW YORK CITY

Work statistics, which measure the proportion of residents commuting to the city for work, provide our strongest indicator of growing regional independence from the central city. According to the 1970 census, approximately 60 percent of the work force residing in Suffolk County also works there. Only 20 percent of the total labor force commutes to New York City. An equal proportion travels to Nassau County, illustrating the interdependence of the region. Suffolk's employment base is roughly comparable to that of the city. New York has 80 percent of its labor force working there with almost the remainder commuting to the Nassau-Suffolk region.

It might be expected that areas nearer to the city would reflect a higher degree of worker commutation. Hence, Nassau County should be more dependent upon New York. Census data, however, reveal that this relationship is not clear. Fifty-one percent of the workers living in Nassau County also work there. However, only 33 percent of the work force commutes to New York with the remainder travelling further out on the Island to Suffolk.

As Figure 1 (Chapter 1) illustrates, some horizontal integration of the suburban submetropolitan region with the central city exists, but not to the degree previously thought to be the case. Sources of employment in Suffolk County have moved it considerably toward independence of New York City. Undoubtedly, both the city and Suffolk are increasingly dependent on national organizations, still, the net balance is such that employment sources are more nearly self-contained than in previous periods. For example, in Privatown, 80 percent of the labor force also works in Suffolk County.[20]

BANKING AND RETAILING: THE GROWTH OF NATIONAL ECONOMIC TIES

By the very nature of suburban growth, which proceeds as new single-family home construction, Suffolk County has become a vast reservoir of liquidity. It represents a huge market for mortgages and loans which have also been insured to a considerable extent by government programs subsidizing the middle class. Consequently, Suffolk represents a highly significant sector of the nation for the realization of banking profits and a source of capital. In fact, Suffolk County has a greater number of banks per number of people than the city. At present, there are 4,800 residents per bank in Suffolk. Queens County, by comparison, which has the highest density of banking in the city of New York, has over 7,000 residents per bank. Of the 36 commercial banks headquartered in Queens, Suffolk and Nassau counties, Suffolk County has 14. The Suffolk banks have 141 branch offices throughout the county and *most* have been built recently. In 1971 Suffolk-based commercial banks had assets of $2 billion. The county also contains the headquarters of seven savings and loan banks. These have 32 branch offices and listed 1971 assets of $725 million.[21] Banking activity by commercial interests headquartered elsewhere are found here as well. While many of these banks have offices in New York City, they are really national and international in scope and hierarchical in form.

As a banking resource the submetropolitan region has its role in the international monetary network. Coincident with the suburbanization process, international banking conglomerates have absorbed small, independent commercial banks which once served the rural population. Marine Midland, Chemical Bank, and Manufacturers Hanover are some of the conglomerates that have bought out local banks. Even Barclary's Bank Ltd. of England attempted to operate in Suffolk, but so far legal constraints have blocked this move.[22]

The suburban social system also plays an important national role as a broad market for retailing. In many areas of Suffolk commercial land-use has doubled and tripled as the suburbanization process proceeds. This development takes the particular and significant form of strip-zoning alongside major, local highways. The proliferation of shopping centers and mile-long strips of commercial stores and services is a basic feature of suburbanization. Such a zoning procedure orders the shopping patterns of consumers; it necessitates the automobile mode of transportation. Thus highway land becomes a major resource of the society, and it is no accident that speculators, politicians, and important *national* financial interests are often involved in the purchase of suburban real estate for anticipated highway use (see Chapters 4 and 5).

The retailing feature of the submetropolitan region is an important part of the suburbanization process. During the decade between 1960 and 1970, Nassau-Suffolk ranked third in the nation in total retail sales. The region outstripped Los Angeles and fell short of only New York City and Chicago. In addition, the region attained the nation's highest expendable income per household during the same period:

> Nassau-Suffolk, like many other suburban communities that have replaced central cities as prime market areas, has greater sales growth

> potential than the nation's top market, New York City. This fact is firmly substantiated in *Sales Management's* preliminary report on its 1971 "Survey of Buying Power," which projects Nassau's growth at 124 and Suffolk's at 124 compared to New York City's 117.[23]

Occupational employment statistics (Table 2.4) suggest the importance of retailing and service industries. Perhaps a better indicator is the changing role of women in the labor force. Between 1960 and 1970 the number of working women increased by 30 percent in Suffolk County. The 1970 figures show that 72.5 percent of these women worked in white-collar retailing jobs, and 15.6 percent in service industries. Only 11.8 percent of the full-time working women were employed in blue-collar capacities.[24] Thus, in the suburban region a large proportion of women are employed on a full-time basis, and probably a similar percentage work part-time. Besides supplementing family income, these women appear to supply a large component of the low-income labor necessary for retailing. Initially, women made the region less dependent than the central city on low-status ethnic laborers by replacing them, but presently, there is an increasing need in Suffolk County for low-wage labor. Privatown in particular, is now the site of the new IRS center and a large state medical complex, which require a growing pool of such workers.

Like the banks, retailing and service sectors have become part of national and international corporate operations. National supermarket chains, gas stations, auto dealerships, shopping malls, and giant department stores dominate the submetropolitan economy. Following the strip-zoned highways we find the corridors of distribution of this social system. The region has also become an operational base for national fast-food chains. The success of this marketing form has been so overwhelming that these food chains are presently being imported into the central city.

An examination of the Suffolk County economy, therefore, indicates the place of the suburban submetropolitan region in the advanced industrial society just as Bensman and Vidich were able to locate the functional niche of the small town.[25] In our case, suburbia is the retailing outlet of the national economy, and its reservoir of liquidity. Corporate suburban activities are important in other ways apart from being components of the national economic system. The expansion of fast-food marketing, for instance, indicates that suburbia is also a region for the development of new national forms of commerce, leisure, and family life.

The 1950s' suburban image of familistic, specialized bedroom communities, suggesting the presence of small specialty shops servicing the limited needs of the population, is much removed from the present submetropolitan economy. Suffolk County is a mass market of families, representing billions of dollars to national retailers and banks. The automobile mode of transportation lies at the core of its marketing process. Control of the Suffolk economy is exercised by locational advantage along strip-zoned highways and the region's 9,746 miles of roads. These commercial zones together with the shopping centers and malls have become increasingly important for major weekly purchases. By contrast, the old town centers have more and more become the sites of small-scale specialty shops, restaurants, stationary and gift stores, sometimes with high locational turnover. Furthermore, it is becoming progressively clear that shopping centers and malls are replacing the old town village square as the organizational focus of community social life.[26] This change in community referrent has some important implications. First, the commercial centers lack the tradition and culture of the older towns, and, therefore, cut people off from reminders of the local past. This effect is heightened by the anarchy of locational decisions, which have contributed to the overconstruction of shopping centers along highways, several of which have been abandoned by merchants and have turned to blight for lack of customers. Second, shopping centers and malls, while serving as meeting places for the community, are primarily commercial enterprises. The informal social needs of consumers often clash with the formal business purpose of such places. One observer, for example, recently remarked that several suburban malls have installed "keep moving" signs, which are enforced by private police and greatly constrain everyday social interaction.[27] Although the older village squares were often patrolled by the local "tinhorn" cop, the same repressive effect on public community discourse was absent.

Thus, there appears to be substantial vertical integration of Suffolk County into the national advanced industrial and commercial system with a decline in the socioeconomic importance of local village life. There is also an increasing interdependence between Suffolk and the rest of the submetropolitan region. In general we assume that each sector is integrated in varying degrees. Nassau County, close to New York City, has become more independent than was envisioned by early conceptions of suburbia. Yet, it is the site of communities such as Great Neck, which seem to fulfill largely a commuter-shed function for very high-income families dependent upon New York City employment. Therefore, both the nearness to New York City and the strength of the national hierarchical marketing forms are factors in suburban regional integration. The structure of this metropolitan pattern is illustrated by Figure 2.

National Corporate and Banking Structure plus Governmental Bureaucracies

Vertical Integration of Subsectors	Urban Subregion		Suburban Subregion	
	Central City	*Zone of transition*	*Suburban sectors*	*Zone of transition*
National				
State	Manhattan,	eastern	Nassau,	eastern
County	part of	Brooklyn	western	Privatown
Township	Brooklyn	and	Suffolk	
Village	and Queens	Queens	and Privatown	
Community				
			Horizontal Integration of Subsectors →	

Figure 2: PATTERNS OF VERTICAL AND HORIZONTAL INTEGRATION OF THE METROPOLITAN REGION OF NEW YORK

Suffolk County appears to perform major functions in retailing, banking, and light industrial production for the national economic system. In turn, the central cities themselves seem to be moving toward a comparable level of specialization. Recently, separate studies by Ganz and by Kasarda have indicated that the city is more the center of coordination and decision-making for the national economy than the locus of coordination for its own local region.[28] Nationally central cities contain a growing recreational and technical service sector catering to the needs of white-collar and professional industries and a large dependent population, who are the clients of the welfare state. Concentrated in the office buildings of cities are the corporate headquarters, government bureaucracies, and centers of mass communication, all of which function to integrate the socioeconomic activities of the entire society.

This view is certainly true of New York City. It contains the headquarters of almost 100 of the "Fortune 500" corporations. It is an international financial center, playing host to the American and New York stock exchanges and the headquarters of most of the nation's investment and commercial banks. New York is also one of the key mass communication and news media centers, which play an increasingly important role in the social integration of activities as U.S. society grows in complexity. The whole state of New Jersey, for example, has no major TV stations, but relies, instead, on those broadcasting from New York City and Philadelphia. Recently, suburban residents have begun to demand media services similar to those in the central city. Cable TV with public access along with UHF broadcasting have partially filled this demand, but cable TV is still largely controlled by networks located in the city. Presently, the role of suburbia in the media is limited. Historically, the city has also upgraded the impoverished populations of the country.[29] That mobility is represented especially by New York City, which continues to perform some of this task for the eastern seaboard.

Suffolk County is not New York City. There are a number of distinct ways in which the submetropolitan region differs from the central city in addition to those we have already discussed. First, it is doubtful that Suffolk County will ever become an important locational site for national corporate headquarters, but a portion of Connecticut is beginning to take on this function. Second, while there has been some cultural decentralization in the areas of mass media and sports, the premiere institutions of the performing arta, the museums, and the yearly cultural festivals, such as in the cinema, opera, and ballet, will remain in New York City. This is so despite the location in the county of a sizable and growing number of little theater groups and museums. They serve to complement rather than replace the massive central-city concentration of cultural activities. A third contrast involves the obvious racial differences in population. New York City's black and Spanish-speaking population is almost 40 percent, while Suffolk is overwhelmingly white (5 percent minority population). In addition to the long history of cultural contributions made by the black community of New York, more recent nonwhite immigration is beginning to enrich the central city and affect the character of life there. Undoubtedly, there is a certain cultural sameness of the submetropolitan region when contrasted with New York City. This lack of diversity, however, is also characteristic of many central cities elsewhere and is not exclusive to suburbia. The relatively more homogeneous submetropolitan population is consequently insulated from the central-city task of having to upgrade the conditions of ethnic and racial minorities. Their welfare burden is considerably less than that of New York City, although it has grown in recent years due to unemployment and the general maturation of the region. Finally, there are important demographic distinctions in age and life-cycle characteristics of the central city versus the suburban population. Suffolk County life styles are still predominantly oriented towards young families and children, although progressively less so. This is reflected in county institutions, which place a great emphasis on recreational activities and maintenance of parks, play areas, and beach fronts, although in part this is done in response to central-city customers. New York City contains a significant proportion of young career-oriented single professionals (called "cosmopolitanites" by Gans)[30] and a large component of elderly people who could not afford migration to other areas of the United States. A commonly felt shortcoming in the submetropolitan region is the inadequate housing, service, and social recognition of these same groups—there seems to be little place for them in the county. This is evident in the housing stock, which still lacks an adequate supply of inexpensive housing for the elderly or for young married couples and housing and social facilities for young professionals. As the population of the county matures, institutional responses may be made in this area, and there are already some indications of change. The tax advantages of having, for example, childless people in Suffolk County are just being realized so that housing for elderly and young people in multifamily dwelling units is increasing. In addition to housing, the limited representation of young marrieds and cosmopolitanites in the submetropolitan population is reflected in the narrowness of cultural activities available to people in Suffolk County. This is not, however, at the minimal level of stultifying uniformity once pictured as characteristic of suburbia. There are, in fact, a growing number of facilities catering to the singles life style in the region. For example, Nassau County has at least two large concert halls, which feature major rock groups, and Suffolk has recently been the construction site for an exclusive singles housing complex and a proposed adult entertainment zone, featuring topless and singles bars and "discos." Nevertheless, the region's cultural and social limitations relative to New York City continue to provide strong incentives for the movement of suburban young professionals into the city and against the middle-class tide.

The combined features of family orientation, affluence, white segrega= tion, limited housing and sociocultural opportunities, and an interdependent economy constitute the relative uniqueness of the submetropolitan region when compared to the central city. Its growing complexity, socioeconomic diversity, housing and land-use problems, and population growth continue to sustain the view that the transformation process begun in the 1950s will continue, but also change.

NOTES

1. Virtually all analysts writing about suburbia in the post World War II decades used the city as a contrasting mode of settlement to organize their discussions. This includes observers who supported the uniqueness of the suburbs, such as Fava, Lundberg, and Donaldson, as well as those that tended to minimize geographical differences, such as Greer, Dobriner and Gans.

2. Cf., Sylvia F. Fava, "Suburbanism as a Way of Life," *American Sociological Review,* 21 (February 1956) pp. 34-37; George Lundberg, et al., *Leisure,* New York: Columbia Univ. Press, 1934; William H. Whyte, Jr., *The Organization Man,* Garden City, N.Y.: Doubleday Anchor, 1956; and A. C. Spectorsky, *The Exurbanites,* Philadelphia: Lippincott, 1955.

3. See the case study of Levittown, Long Island, in William Dobriner, *Class in Suburbia,* Englewood Cliffs, N.J.: Prentice-Hall, 1963; and Port Washington in L. Sobin, *Dynamics of Community Change,* Port Washington, N.Y.: Ira J. Friedman, 1968.

4. Bennet Berger, *Working-Class Suburb,* Berkeley: Univ. of California Press, 1960; William Dobriner, op. cit.; and (ed.), *The Suburban Community,* New York: Columbia Univ. Press, 1969; Herbert J. Gans, *The Levittowners,* New York: Pantheon, 1967; David Thorns, *Suburbia,* London: MacGibbon & Kee, 1972; Leo F. Schnore, *The Urban Scene,* New York: Free Press, 1965; and with Joy K.O. Jones, "The Evolution of City-Suburban Types in the Course of a Decade," *Urban Affairs Quarterly,* 4 (June 1969), pp. 421-443.

5. Scott Greer, *The Emerging City: Myth and Reality,* New York: Free Press, 1962; Jean Gottmann, *Megalopolis: The Urbanized Northeastern Seaboard of the United States,* New York: Twentieth Century Fund, 1961; and with Robert A. Harper (eds.), *Metropolis on the Move,* New York: John Wiley, 1967; Paul D. Spreiregen (ed.), *The Modern Metropolis: Selected Essays by Hans Blumenfeld,* Cambridge, Mass.: MIT Press, 1967.

6. Despite the importance of the suburban submetropolitan region there is still no consensus on the definition of these areas. The Regional Plan Association suggests that suburban regions can be defined in terms of persons per square mile, and that densities between 1,000 and 10,000 people represent "sprawl," cf., Regional Plan Association, *Growth and Settlement,* New York: 1975.

7. *Newsday Sunday Magazine,* 4/29/73, p. 9.

8. *U.S. Census of Population,* 1960 and 1970.

9. Regional Planning News, 91, (September 1969). This process is still continuing. Between 1970 and 1975 New York City's white population declined by an additional 600,000. Presently, the city is 62 percent white compared with 87 percent white in 1950.

10. *Newsday Sunday Magazine,* op. cit., p. 8.

11. *Newsday,* "The Moving Motive: A Better Life," (4/30/73), p. 4.

12. James R. Hudson, "Surveying Community Attitudes," in Dieter K. Zschock, (ed.), *Brookhaven in Transition: Studies of Town Planning Issues,* Stony Brook, N.Y.: SUNY at Stony Brook, 1968, p. 52.

13. Ibid., p. 50.

14. Ibid., p. 52.

15. The mature subregion of Nassau County is already attempting to face the many consequences of decline in population growth after years of expansion. See for example, Larry Levy, "No-Growth: A New Era of Less-is-More," New York *Times Sunday–Real Estate Section,* 12/7/75, p. 1. Some of its problems are: surplus public buildings, deficit spending, and shopping center blight.

16. *Newsday Sunday Magazine,* op. cit., p. 8.

17. For a general discussion of this trend cf., Robert Schafer, *The Suburbanization of Multi-Family Housing,* Lexington, Mass.: Lexington, 1974.

18. *Newsday,* 3/22/71, p. 2R.

19. Ibid., p. 2R, 3R.

20. Our census figures for Nassau and Suffolk counties as well as Privatown seem to be comparable to the national experience. For example, a study done in the 15 largest metropolitan areas found the proportion of suburbanites who both live and work in suburbs to be 72 percent. Cf., Jack Rosenthal, New York *Sunday Times,* 11/15/72, pp. 1, 58.

21. Long Island *Commercial Revue,* 7/26/71, p. 3.

22. Cf., Terry Robords, "How New York Banks Have Spread Out," New York *Times, Sunday, Business and Finance,* 11/2/75, section 3, p. 1.

23. "Marine-Midland Survey," Long Island *Commercial Review,* 1/26/71.

24. *U.S. Census of Population–occupational characteristics 1960, 1970.*

25. Joseph Bensman and Arthur J. Vidich, *Small Town in Mass Society,* Princeton, N.J.: Princeton Univ. Press, 1958.

26. Cf., Rose De Wolf, "Main Street Goes Private," *The Nation,* 12/18/72, p. 626.

27. Leonard Downie, Jr., *Mortgage on America,* New York: Praeger, 1974, p. 103.

28. Alexander Ganz, "The City: Sandbox, Reservation or Dynamo?" *Public Policy* 21 (Winter, 1973), pp. 107-123; John Kasarda, "The Theory of Ecological Expansion: An Empirical Test," *Social Forces,* 51 (December 1972), pp. 165-175.

29. Ganz, ibid.

30. Herbert J. Gans, "Urbanism and Suburbanism as Ways of Life: A Re-evaluation of Definitions," in Arnold Rose (ed.), *Human Behavior and Social Processes,* 1962, pp. 625-648. Gans views metropolitan community differences mainly in terms of social class and life-cycle stage rather than residential location per se, as does Greer and Dobriner. We tend to agree with this view.

Chapter 3

THE SOCIAL ORDER OF PRIVATOWN

Privatown contains the suburban fringe area of metropolitan expansion. During the past decade, it has experienced the most rapid population growth on Long Island, increasing by 121 percent between the 1960 and 1970 censuses.[1] The five townships located further east still represent rural-recreational communities. They include the Hamptons and Shelter Island as well as the rich farmland of the east end, such as in Southold. The area to the west, including Privatown, however, has already experienced suburban growth and is maturing in the same manner as Nassau County.

Although the most recent area of development, Privatown is not characterized by suburban homogeneity. To be sure, the composition of particular communities within the township may be relatively uniform, but the area as a whole is stratified by income, housing use, race and lifestyle. This same pattern of relative homogeneity within a community, on the one hand, and the aggregate diversity of the entire region, on the other, has always been a commonplace observation of the city itself. In addition, the progressive stratification of the area has had some effects which are also comparable to central city patterns. There are increasing inequities in social service supply, racial and income segregation, high tax rates and large differences in assessment, and significant deterioration and blight of some communities. In short, Privatown has emerging within its borders a pattern of local differentiation with the attendant inequities that are familiar from studies of our central cities.[2] While we cannot predict that such a process will reach a level which will overtax the limited resources of the township, we can, nevertheless, be fairly confident that the coming years will see the suburban regions echoing the very same calls for federal assistance as are presently being voiced by city governments. If so, the disappearance of the homogeneous, untroubled suburbia has some very important implications beyond the abandonment of invalid premises regarding earlier conceptions of its social composition. As we shall see below, the county is already at the stage of requiring extensive federal aid in order to complete its sewer construction program.

The discovery of this diversity raises the question—what is causing it? Are the factors which have created urban-style heterogeneity cultural or interactive in nature and, therefore, endemic to the society as a whole? Or, alternatively, do certain demographic or structural variables account for the submetropolitan settlement pattern? In Suffolk County it appears that the structural variable of short-sighted land-use policies can largely account for the observed trends. Our discussion of Privatown's social order will first attempt to demonstrate this conclusion. We will then look at the social problems of suburban diversity, and lastly examine their implications.

Ecological Adjustment: Housing and Income Effects

New York metropolitan expansion was stimulated by planning programs at the federal level, including the FHA housing legislation and the Interstate Highway Act. These did neither anticipate the parochialism of local areas, nor the limited impact of bureaucratic planning. Lacking a dominant governmental unit that could effectively control growth, submetropolitan development has been largely uncoordinated. This lack of integration produced ecological processes of adjustment in response to rapid demographic shifts. Between 1950 and 1970, residents in Privatown desired to discourage further development of the area. In the absence of effective programs designed to aid local-level planning, and as a consequence of the limited vision of the local political organization, the residents themselves called upon the township government to implement restrictive selective growth strategies. The town board's response was

Table 3.1: Up-Zonings of Residential Land in Privatown, 1950 to 1970

Type of Zoning Single-Family Tracts	Amounts of Acreage in Category		
	1950	1962	1970
1/1 acre	--	--	241
3/4 acre	--	4,630	13,275
1/2 acre	--	244	39,926
1/3 acre	--	34,058	70,591
1/4 acre	3,000	--	2,560
1/5 acre	--	--	--
1/6 acre	23,000	53	--
1/8 acre	113,000	1,120	--

NOTE: These data from township maps, were obtained by Richard Camitta in conjunction with a previous study. Cf., Richard Camitta, **A Re-Examination of the Racial Contact Hypothesis,** Ph.D. dissertation, Department of Psychology, Stony Brook, N.Y.: SUNY at Stony Brook, 1973.

characteristic of other local governments on Long Island. The residents obtained up-zonings over a twenty year period on all residential land in the township to larger tracts.[3] This was accomplished by having the town board revise the zoning categories of residential land to require greater acreage per residential lot. The process was accomplished most successfully by the wealthiest areas, especially those along the north shore. They wielded the most political influence, and their style of life was most threatened by small-lot developments. Table 3.1 illustrates the shift to larger tracts, called "up-zoning," in Privatown.

Up-zoning works in conjunction with the zoning code regulating house construction on each plot of land. It has the following effects: First, by requiring that all new housing be built on larger plots, it helps to limit the future density of development. This is coupled with distinct provisions against multifamily dwellings in most areas to retard population growth. Second, the larger plots come attached with specific zoning and building code regulations, which require that new housing have a relatively large minimum size. Such constraints raise significantly the costs of construction so that only the wealthier migrants can afford to own homes in the area. The costs rise because minimum size requirements affect the amount of land needed for building, the frontage necessary to be developed by the builders, and the extra construction costs of the house itself. This constrains and limits the filter effect of new housing. With less total housing available and with the costs of individual new homes rising, the value of all housing in the township appreciates significantly even if its quality does not improve. As elsewhere, behind the apparent "free" action of supply and demand there are important constraining politically inspired regulations.

The practice of local government to control growth in this limited fashion has been called "exclusionary zoning." Analysts of up-zoning activity have indicated that because all zoning is exclusionary it is perhaps redundant to reserve the concept specifically for this phenomenon.[4] Nevertheless, such practices are exclusionary in a particular way because they specifically and primarily limit the ability of lower-income groups, especially minorities, to live in most suburban communities, and this fosters income and racial segregation on a metropolitan scale. In fact, "exclusionary zoning" has emerged as a key social issue of the nation.[5] Because of the congressional failure to initiate national land-use legislation, civil libertarians have called upon the courts to determine equitable regional growth policies. As in the case with the entire issue of zoning and private property rights, local judicial rulings have been inconsistent. For example, a New Jersey state supreme court decision against Mount Laurel township in March 1975 declared exclusionary practices unconstitutional. It ruled that townships must consider regional as well as local housing needs and should strive to maintain greater economic and racial balance.[6] On the other hand, in August 1975, a federal appeals court in California ruled in favor of an exurban village, Petaluma, in the San Francisco bay-area region, and supported its desire to restrict growth through exclusionary zoning. The ruling stated that ". . . it was a valid exercise of Petaluma's zoning power to limit growth to preserve its small-town character," and said it was not up to the federal courts to act as a "super-legislature of a zoning board of appeal."[7] Such inconsistencies may be resolved with a U.S. Supreme Court ruling for which Petaluma may well become the test case.[8] In any event, the present confusion, occurring in the absence of clear guidelines for regional versus local land-use needs, operates to the advantage of local interests in Suffolk County. Any forthcoming rulings by the courts on exclusionary zoning will, therefore, greatly affect the entire county and especially its less-developed eastern end.

The zoning policies of suburban townships have been attacked by civil libertarians as restricting the ability of Blacks and poor to find housing in a region. An examination of Privatown, however, reveals that the effect of up-zoning residential land has increased segregation among *all* the people in the township. These effects, therefore, *go beyond* the present-day conception of exclusionary zoning. Over the years, a socioeconomic gradient has emerged as operating in the settlement pattern of Privatown among more or less affluent areas. Communities differ substantially by median value of home and family income and these differences are directly related to their location within the township itself. This pattern is a function of zoning because zoning and construction practices develop different areas for different income groups, largely depending upon the residential land-use category and its prescribed house plot.[9] Developers in Suffolk County operate on vacant land and build many single family homes and the necessary infrastructure of streets and roads all at once. The resulting community is an enacted one.[10] The tract-housing mode of construction builds entire communities within a given region, zoned single family residential, and the homes are all roughly valued at the same price. Buyers arrive almost conjointly and take possession from the developer not only of their homes, but of the name of the community, and, in addition, of a superficial articulation of its character and form. Constructed all within the same residential tract and with a similar style, the new suburban community taps into a specialized segment of the housing market and presents purchasers with an embryonic community structure. At the outset, the suburban development is homogeneous in form and in its socioeconomic character. Heterogeneity between communities within the township then arises because of the presence of variety in the residential categories of the zoning ordinance extending from one acre to as little as one-fifth of an acre, and in some locations allowing for multifamily dwellings. This results in a significant mix of housing costs by location within the township and creates the condition for the aggregate diversity in land-use observed by suburban analysts. The relatively homogeneous, enacted community tailored to the specific zoning category constitutes the building block of the suburban residential array.

In Privatown the location gradient runs north to south. On the north shore are located the relatively wealthy communities, some of which have grown around a nucleus of upper-class estates and incorporated villages. This strip contains developments with very low-density tracts and is the site of much up-zoning activity and other exclusionary practices.[11] The amenities of the water front encourages this usage along with the relative power of the residents. The south shore is similar in tenor, yet it is not as exclusive nor as homogeneous as the north. Developed much earlier as small, somewhat independent communities this subregion contains a number of working-class housing areas including the incorporated village of Patchogue which was an early industrial location on Long Island.

Through the center of the island runs the transportation and industrial spine. It is the site of the Long Island Expressway as well as the Long Island Railroad's mainline right of way. It is here, furthest from either shore, that the least expensive housing is available. This locational gradient is illustrated by Table 3.2, which indicates median family incomes in Privatown by community.

Community heterogeneity exists because of the family income diversity of the mobile suburban population. Along with upper middle-class professionals moving to better housing in Privatown, there are many blue-collar and lower- middle-class families who have been driven out of Nassau County

Table 3.2: Median Family Incomes in Privatown by Community and Location for 1950, 1960, and 1970

	1950[d]	1960	1970	Change 1960-1970
South Shore				
Bellport	NA	$8,200	$12,514	+4,314
East Patchogue	$3,261	6,564[a]	10,868	+4,304
Patchogue	2,917	6,115	9,547	+3,432
Mastic Beach[b]	NA	4,810	8,110	+3,300
Center Moriches	NA	5,930	10,446	+4,516
Center				
Centereach		$6,474	$10,997	+4,523
Lake Grove		NA	11,562	– –
Shirley[b]		4,900	9,137	+4,237
North Bellport[c]		5,881[a]	8,429	+2,548
North Patchogue		5,776[a]	10,381	+4,605
Ronkonkoma		6,196	10,686	+4,490
Yaphank[b]		5,502[a]	9,973	+4,471
Holbrook-Holtsville		6,206	10,788	+4,582
Selden		5,651[a]	10,581	+4,930
North Shore				
Pt. Jeff. Station		NA	$11,117	– –
Pt. Jefferson	$2,953	$6,995[a]	13,129	+6,134
Setauket		8,077[a]	14,664	+6,587
Stony Brook		8,890	17,083	+8,193
Rocky Point-Shoreham[b]		6,434	10,528	+4,094
Difference Between High and Low	$344	$4,080	$ 8,973	+4,893

NOTE: U.S. Census of Population, 1950, 1960, and 1970.
a. Data approximated from census tract.
b. Located in the eastern (rural) section of the town.
c. Major black ghetto (40 percent black as against 2.6 percent for town).
d. The column shows all that is available for communities from the census of that year. This is mainly due to the relatively unpopulated state of these areas at that time.

or towns closer to New York City as a consequence of rising taxes and cost of living. No doubt some comparatively wealthy migrants buy homes in the center, such as in the community of Selden, but those who can afford it normally prefer housing near the waterfront.

Several effects of the suburban transformation are illustrated by Table 3.2. First, between 1960 and 1970, differences in family income, almost equal in 1950, have more than doubled between the poorest and most affluent communities. Second, the poorest communities in each of the three township regions are those located furthest east. They are presently beyond the advancing effects of the transformation wave.

This income gradient is most pronounced when the median value of housing is examined. As a consequence of rapid development and up-zonings the wealthiest communities have experienced the greatest appreciation of the value of their homes. Defensive strategies carry a dividend in suburban growth because as demand for housing rises, the more exclusive communities become progressively the more desirable places to live. The value differential for housing by location is illustrated in Table 3.3. Small differences in 1950, as in family income above, have become highly significant by 1970.

The Long Island daily newspaper, *Newsday,* commented on the importance of the trend represented by Table 3.3. It observed the same effect operating in the rest of the submetropolitan region.

Table 3.3: Median Dollar Value of Homes by Location in Privatown, North and South Shore vs. Center, and Difference in the Distribution of Value

	1950	1970	Change
North and South Shore			
East Patchogue	$ 8,264	$19,280	$11,061
Patchogue	8,064	18,805	10,741
Port Jefferson	9,293	27,655	18,362
Center Moriches	8,123	20,755	12,632
Center			
Centereach	$10,914	$19,765	$ 8,851
Mastic-Shirley	6,490	12,110	5,620
North Bellport	8,129	12,095	3,966
Ronkonkoma	9,519	19,615	10,090
Selden	6,793	21,050	14,257
Difference Between High and Low	$ 4,121	$15,560	$14,396

NOTE: See Table 3.2.

> The median value of homes on L.I. has skyrocketed by more than 60% in the last ten years (1960-1970). But there is a widening gap between communities of lower-priced homes and the rising median, a trend which housing experts fear may lead to the deterioration of the Island's less affluent communities.
>
> Wealthy areas have continued to climb rapidly in value, so there are now 46 L.I. communities with a reported median house value of more than $50,000. At the same time, however, poorer communities in both counties have fallen further away, in some cases substantially, from the two county medians since 1960. The statistics indicate a dramatically widening gap between the very wealthy and the less affluent areas.[12]

As might be expected the locational gradient creates differences in occupational backgrounds for submetropolitan communities. Taking work categories as a reasonable indicator of class/status differences, we can see the extent of segregation by location for Privatown in Table 3.4. Such variation by community is also associated with differences in family income as indicated in Table 3.2. Together, these data suggest significant contrasts in life style for suburban developments. One way this can be seen, involves the contrasting attitudes expressed by residents whose occupational background and family income differ with regard to their views on community life, family, and social issues. A 1973 survey of Suffolk County communities revealed that residents of affluent developments tended to be more satisfied with their local community life and housing, more interested in supporting social measures against racism, and more tolerant of the behavior of young people, than were residents in lower-income suburban areas.[13]

Field-work investigations in Privatown support the existence of such life-style differences by community location, although developments can

Table 3.4: Percentage Occupational Breakdown by Residential Location, North and South Shore vs. Center in Privatown for 1970

	North and South Shore		Center	
Percentage of Total Professional Employment	61%	(9)	39%	(11)
Percentage of Total Blue-Collar Employment	34%	(9)	66%	(11)

NOTE: See Table 3.2.

scarcely be considered totally homogeneous. Many shore communities can be characterized as populated by "organization-man" professionals, and some, such as "Executive Estates," attempt to attract this group. Center communities tend to be middle- or working-class with everyday activities organized around strip-zoned commercial highways and higher-density residential land-use. Shore communities, in contrast, are better landscaped and leisure oriented. The north shore, for example, tends to have many parks, golf courses and an extensive beach area, fronting the Long Island Sound, which is used extensively by the communities located there. The south shore developments are also more affluent than the center, but do not have direct access to beach front. In this section residents must cross the Great South Bay to public areas on Fire Island in order to enjoy this well-known and attractive ocean recreational region, which extends for miles. Boating, fishing, swimming and clamming, therefore, are all activities directly associated with the Long Island life style. Residents located on the shores often have direct waterfront access from the backyards of their homes. The communities located inland provide no such access to recreational activities; their residents must commute and then use the limited public facilities for township residents, such as public beaches or marinas, in order to enjoy the region's amenities, much like people coming from New York City.

Life-style differentiation and segregation are outgrowths of defensive strategies and postures that are related to a common desire, but differential ability, to protect the community against the advancing wave of metropolitan expansion. In Suffolk County these sentiments, shared by many of those being excluded incidentally, support the legal and administrative measures which control home construction by local government and regulate the type of house and the materials used in building. The residents' desires to preserve the value of their property by regulating land-use and construction in the face of uncoordinated growth complement the local political organization's desire to harness rapid real-estate development to support itself. These efforts have combined to increase the price of all housing in the region, to constrain greatly the architectural variety of design, and to segregate developments by race, income, and life-style.[14] The local submetropolitan zoning and building practices, moreover, have arisen from a need to adjust to seemingly endless suburban development and to the failure on the part of both national and local government to coordinate social activities in the broad public interest. A long-range vision of what regional growth should be like is generally exercised only by the planners and the local religious and housing reform associations interested in social equity. Both the residents and the local political leadership have been reluctant for different reasons to participate in such long-range visions. Due to the failure of land-use planning and the weakness of local government, however, it is progressively less certain that resident desires to preserve property values, recreational enjoyment, privatized family life, educational quality, and tax levels will be protected in the future.

The presence of substantial constraints on the building of homes in Privatown has further implications. First, as we have indicated, these constraints have contributed to the bureaucraticization and the politicization of the housing industry. Second, one of the consequences of defensive strategies is the rise of urban social problems in Privatown. These are seen to be a result of the socioeconomic locational gradient and their presence compels us to reexamine the efficacy of such parochial patterns and attitudes.

Social Problems of Privatown

RACIAL EFFECTS

Up-zonings, by themselves, cannot explain the regional racial homogeneity of suburban communities. As in the central city, defensive strategies have been devised primarily to protect racial uniformity. Developers selling new housing must be in a position to assure prospective customers that minority purchases of homes will be limited, and their task is made easy by the high incomes required. Second, real estate agents act as gatekeepers to filter out not only blacks, but other possible "undesirables," such as communal groups, unmarried couples, gays, and interracial couples.[15]

Black people in Privatown live in suburban "ghettos." The township has a nonwhite population of only 3 percent, mostly confined to three areas, Gordon Heights, with a 1970 minority concentration of 46 percent, North Bellport, which has 40 percent minority groups, and Mastic Moriches with 20 percent. These areas are the "new" suburban black ghettos and together they include almost the entire black population of the township. Important to note, however, is the fact that they did not become ghettos as a consequence of suburbanization or exclusionary zoning, which merely exacerbated the already existing racial segregation. For over 50 years, these communities have been the only areas where blacks could be moderately successful at finding housing. A summary of the history of Gordon Heights indicates this:

> This area was the creation of a real estate broker in New York who in 1927 recognized a market among Negroes for property on Long Island. He sold plots of land to Negroes who built their homes on the parcels. More recently, a developer built homes in the Coram section of Gordon Heights and sold them to Negroes. This is one of the very few locations in the town of Privatown where a Negro can buy a house. There, it has become a colored ghetto where more than 1,500 people reside.[16]

The process of suburbanization abetted racial segregation of the region in two ways: (1) It reinforced the ghettoization of Blacks, and (2) it fostered the deterioration of the areas through the emergent features of the income differential. A study done by the bi-county planning board of Nassau-Suffolk found that in the past ten years, Long Island's black population have continued to move into the areas where blacks already lived. "It's not a question of some of these areas becoming ghettos," said the head of the board in a newspaper interview, "they already are."

Outside the central city such housing patterns are likely to segregate minority school-age children despite their small numbers. Suburban ghettos have racially segregated schools with no avenue to appeal the discriminatory residential patterns which caused them, because there is no centralized municipality responsible for maintaining racial balance.[17] In Privatown schools are controlled by the community or the school district, and not the township.

Another effect of racial segregation is the deterioration of housing in ghetto areas and the creation of blight. Table 3.3 indicates that the community of Mastic-Shirley experienced a relatively small increase in the median value of housing compared to the substantial increase of other areas on the north and south shore. North Bellport's increase was even less. In fact, if inflation is taken into consideration, some of these homes may actually have depreciated in value. Housing studies of the township support this contention. As early as 1960, 31 percent of the available housing in Gordon Heights and 20 percent in Mastic-Moriches was classified as substandard.[18]

INCOME EFFECTS

Income segregation on the scale of the entire suburban submetropolitan region creates a number of problems for socialization. One effect is on education. There is a direct association between location of school and the district competency level. An examination of Privatown school districts indicated that the wealthier schools located on the north shore have higher competency rates than schools located elsewhere in the township, especially in the less affluent and working-class oriented districts. The same geographical gradient observed in family income, housing, and lifestyle differences by community location also appears to be operating with respect to pupil competency. These effects are illustrated in Table 3.5.

The supply of social services is also affected by income segregation. Municipalities often exhibit a pattern of differential supply of publicly supported services as a function of income. It has been claimed that such supply differentials occur in New York City in the services of fire and police protection and in garbage collection. Whether or not this is true of County supported services, an analysis of Privatown reveals an analogous pattern in social-service supply. Medical care, for example, is almost entirely concentrated in the wealthier north-shore communities. Eighty-two percent of all doctors have offices located on the north and south shore with 57 percent located in the "three village" region of the north shore

Table 3.5: Percentage of Students Below Minimum Competence in Testing Scores by School District and Location, 1970

	% Below Minimum in		Approximate Median Family Income, 1970
	Reading	Math	
North Shore			
Port Jefferson	13	16	$13,129
Port Jefferson Station	23	22	11,117
Three Village	14	13	+14,000
Rocky Point	13	18	10,528
Center			
Bellport (North)	34	33	$ 8,429
Middle Country	27	27	+10,000
Middle Island	not available		
William Floyd	not available		
Sachem	not available		
South Shore			
Patchogue	29	30	$ 9,547
South Manor	18	18	– –
Center Moriches	21	18	10,446
East Moriches	12	12	– –
Mastic Beach	35	34	8,110

NOTE: A. Bruce Borass, **Education** (unpublished paper), Stony Brook, N.Y.: SUNY at Stony Brook, Applied Ecology Project, 1973.

as against population concentration. In addition, all of Privatown's four hospitals are located in the shore areas, with three located in the same "three village" region. Recently a major university hospital was located in the same area as well. Thus, there is virtually no medical care conveniently available for the poorer residents in the center of the island and they must travel to either shore to obtain treatment. In this case, the pattern of supply has some further consequences. For example, the Long Island Expressway, the major transportation artery of the island, runs through the center, and automobile accident injuries there have an extremely high probability of resulting in death. Victims who might be saved through prompter care die due to the time it takes medical aid to reach the center. Suffolk County has led the state in auto fatalities each year since the 1960s. A report in *Newsday* on this phenomenon stated:

> Emergency medical care in Suffolk County today is so fragmented and obsolete that I thank God when I drive back over the Nassau line. If you're hurt in an auto accident in Nassau, your chances of survival are nearly twice as good as they would be in Suffolk. In fact, if you have a heart attack or any kind of serious accident, you're better off in Nassau. Under present conditions, I wouldn't want to live in Suffolk . . . A study of some 3,000 runs by Suffolk ambulance units indicated that the average time from accident scene to hospital ranged from 26 to 33 minutes. This is considered much too long to save lives.[19]

In Privatown, therefore, life style, minority-group locations, educational opportunity, and availability of medical care are highly segregated despite the relatively high level of income in the region. There is a general drift toward the aggregation of these social inequities created by residential segregation which belies the formerly perceived cultural sameness of suburbia and which progressively resembles the social patterns of the central city.

SPREAD CITY

Perhaps the clearest manifestation of the high social costs of suburban land-use patterns can be seen in its present form of development. Besides helping to create an income gradient in the area, up-zonings have helped to stimulate the massive expansion of the metropolitan region. Low-density tracts contribute to suburban sprawl, the primary anathema of planners. The Regional Plan Association, a private agency interested in combating the adverse effects of metropolitan growth, announced over a decade ago that low-density zoning was the most inefficient use of social resources. Such a pattern leads to the private usage of most open space and makes it difficult for suburban municipalities to plan adequately for social services such as mass transportation. In a 1962 analysis, the association had already called attention to the future costs of such growth.

> In the spread city, decreed by present zoning, people will be living and working too far from each other to use public transportation or to walk to most places they want to go, or even to car-pool. This adds to the spread by increasing the roads and highways needed. It also limits everyone to one mode of transportation and increases the time and cost of bringing people together. . . .
>
> Costs of spread-city, especially for transportation, will be much higher than costs making full use of the older cities. Municipalities will be more than ever inclined to indulge in "fiscal zoning" trying to zone out tax users (families with children) and zone in tax providers (industry). Tax considerations, in short, will play an expanding role in land development decisions, weakening the chance of planning for the best possible use of the land.[20]

This warning went unheeded by the local political organizations in control of township zoning. They catered instead to the narrow interests expressed by the selective calls for up-zoning. Presently, the entire suburban submetropolitan region faces the social costs of uncoordinated spread-city growth. These are reflected in a number of concrete ways. First, such a pattern indirectly constrains future attempts at guiding land-use in a more planful fashion. The Nassau-Suffolk Bi-County Planning Board, the regional authority entrusted with the master planning function, has had only very limited success over the past decade in controlling land-use and in affecting local zoning policies.

Second, the pattern has directly resulted in the need for tax supported programs to counter the inequitable social costs of low-density development. The decline in open space, for example, has forced Privatown to use public funds to purchase land for parks at prices competitive with residential use, in many cases $5,000 per acre. Like central-city residents, individuals not living directly on the shores of Privatown must commute to recreational areas and still pay taxes for their upkeep. Perhaps the clearest indicator of the social costs of such a developmental pattern is, in fact, illustrated by tax rates. In 1970, Privatown residents paid combined state, county, and township property-tax rates ranging from $108.41 to

$126.08 per $1,000 assessed value. These rates have increased steadily each year and are expected to grow in the future. In 1970, the average Suffolk family paid $1,500 in local property taxes. Nassau County had the fifth highest rates in the nation, with an average of more than $2,000 per family. High taxes is the major source of resident discontent in the region. A 1973 survey of attitudes towards suburban problems, for example, revealed that 68 percent of the Suffolk citizens sampled were primarily concerned with the high costs of keeping their homes. This was the first concern of respondents in every category of income, race, and place of residence.[21] Added to this growing property tax burden are the rates which support the local school districts. In Privatown these vary greatly by community, ranging in 1971 from a low of 11.23 per $1,000 of assessed value for Port Jefferson to a high of 24.59 for Port Jefferson Station. The wide differences and inequitable tax burdens carried by communities that are often located next to each other is a result of decentralized school financing. Revenues raised in support of education depend upon the ability of each district's industrial, commercial, and residential property values to generate taxes, and this varies greatly within the township. Along with other property taxes the burden on homeowners in the region is presently extreme compared to the remainder of the nation, and school districts with limited tax generating abilities have had to cut back drastically in support for educational programs. The present fiscal crunch in suburban education is analogous to the problems encountered in central city districts.

The disappearance of open space is a consequence of the wholesale environmental negligence of spread city. The environment of any region depends upon the natural processes located in some open, undeveloped areas. These provide recharge basins for the water table, reproduction sites for birds and wildlife, oxygenating areas as a hedge against air pollution, and recreational sites for people. A lack of balance between such areas and the concrete-layered land of highways and housing extending throughout the metropolitan region can upset the local ecology and lead to environmental problems. Suffolk County residents are extremely concerned with the environmental effects of water, air and shore pollution. Recently, sewage has been added to this list. Because county regulations did not require developers to provide effluent treatment other than backyard cesspools, the Suffolk communities have no sewers. The present level of development has caused an "effluent" crisis. There is a lack of adequate treatment of sewage and inadequate recharge of the water table. To meet this crisis local municipalities have had to levy additional taxes on the residents, and a county-wide bureaucracy has been established which recently embarked on a sewage-construction program supported by fiscal bonds. Inadequate planning by government has thus necessitated the creation of another bureaucracy, deficit spending, and higher individual taxes. Like other bureaus of government, the Suffolk County Department of Environmental Protection has been plagued by programs that have met with only limited success and resulted in increased taxpayer ire. In fact, the sewage construction program has already overtaxed the county's ability to support it. It is presently proceeding with the aid of state and federal funds.[22]

The overriding lesson of metropolitan expansion is that the present pattern of growth preserves and worsens the social inequities of the local area, which have become ubiquitous problems of social control, have led to deficit spending by the county, and the need for outside financial aid. Ultimately, it is the local municipality which must carry the burden of inadequate planning, and it is the local citizen who will be called upon to pay for its social costs. Unlike the city, however, the suburban townships, such as Privatown, have neither the resources nor the local governmental structure to adequately confront the broader inequities of societal patterns of development. There is a growing inability to share the social costs of growth and to aggregate public opinion across the region. The irony of this situation will probably come to haunt the privately-oriented population in the future. Holding fiercely to their belief in the autonomy of local government, suburban residents are watching the growing erosion of local power at the hands of federal and state bureaucracies.

NOTES

1. Privatown is larger in area than Nassau County, covers almost one-third of Suffolk County and is the fifth largest township in the state. In 1972, 40 percent of the land was undeveloped, and Privatown is expected to be the site of Long Island's greatest growth for the next two decades. The 1970 population of 267,245 is expected to double by 1985. Cf., Tom Morris, in *Newsday*, 6/23/72, p. 4.

2. For a general discussion of these maturing patterns, see, Jeffrey K. Hadden and Louis Masotti, (eds.), *The Urbanization of the Suburbs,* Beverly Hills: Sage, 1973. See also Hadden and Masotti, (eds.), *Suburbia in Transition,* Chicago: Quadrangle, 1974.

3. To avoid a point of confusion–please note that these changes are called downzonings in some areas.

4. Bernard Siegan, *Land Use Without Zoning,* Lexington, Mass.: D. C. Heath, 1972. Siegan's case study of Houston, Texas contends that free market incentives might create a more desirable land-use pattern than government zoning restrictions and would reduce housing costs. For a discussion of this study's limitations, cf., Richard Babcock and Fred Bosselman, *Exclusionary Zoning,* New York: Praeger, 1973, p. 89. For a general discussion of the ways in which all communities have been using land-use controls to respond to uncoordinated growth, cf., Franklin J. James, Jr. and Oliver Duane Windsor, "Fiscal Zoning, Fiscal Reform and Exclusionary Land-Use Controls," *AIP Journal,* (April 1976), pp. 130-141.

5. Babcock, op. cit. For additional statements on exclusionary land-use as a social control mechanism, cf., John Delafons, *Land-Use Controls in the U.S.,* Cambridge, Mass.: MIT Press, 1969; Paul Davidoff and Neil Gold, "Exclusionary Zoning," *Yale Review of Law and Social Action,* 1, (Winter, 1970), pp. 58-60; Syracuse Law Review, 475 (1971); Marion Clawson, *Suburban Land Conversion in the U.S.,* Baltimore: Johns Hopkins, 1971.

6. *Daily News,* 3/25/75, p. 16. In October, 1975, the U.S. Supreme Court, following its pattern of leaving land-use decisions to local areas, upheld the state ruling.

7. New York *Times,* 8/17/75, p. 34. These seem to be the sentiments of the U.S. Supreme Court as well.

8. Judging from the past, the Supreme Court does not wish to make national land-use policy, and legal inconsistencies will no doubt persist in the future.

9. This is one reason why single communities may appear to be homogeneous. Suburban heterogeneity exists on a regional basis, and early studies have failed to appreciate this aspect of submetropolitan land-use, especially when they select only a single community to do a case study.

10. Cf., Gerald D. Suttles, *The Social Construction of Communities,* Chicago: Univ. of Chicago Press, 1972, for a discussion of the general properties of enacted communities.

11. In 1974 the U.S. Supreme Court ruled on a Privatown case involving the incorporated village of Belle Terre. The court upheld a village ordinance against "groupers," the renting of a home by three or more unrelated individuals. Anti-grouper ordinances have proliferated since this ruling across the island and have even adversely affected the Southampton summer rental trade.

12. *Newsday,* 2/18/72, p. 5.

13. Ibid., 5/4/73, p. 8.

14. Obviously, there are analysts who do not believe such a community pattern and land-use constitute a social problem. One argument suggests local autonomy as a way of distributing public goods proportionately to those who can afford it, rather than making equitable distribution the responsibility of the entire region or society. An early statement of this position is Charles Teibout, "A Pure Theory of Local Expenditure," *The Journal of Political Economy,* (1956); and a more recent formulation of this argument relevant to exclusionary zoning is Edwin S. Mills and Wallace E. Oates, *Fiscal Zoning and Land Use Controls,* Princeton, N.J.: Princeton Univ. Press, 1975.

Our study of Long Island, in contrast, demonstrates that heterogeneity on a massive, regional scale produces serious inequities and public good distribution problems. Federal investigations of urban problems also seem to support our view: cf., *Report of the National Commission on Urban Problems* (The Douglas Commission), Washington, D.C.: U.S. Government Printing Office, 1968.

15. Cf., note 12 above.

16. *Housing Problems in Brookhaven,* Suffolk County: League of Women Voters, 1965, p. 15.

17. For a full discussion of these effects, cf., Camitta, op. cit.

18. *Housing Problems in Brookhaven,* op. cit.

19. Mathew Bonera, "Why They Die in Suffolk," *Newsday,* 6/22/72.

20. *Spread City,* New York: The Regional Plan Association, 1962, pp. 2-3.

21. *Newsday,* 5/5/73, p. 8.

22. For a case study of the growing effluent problem and its fiscal results in Suffolk, see the following newspaper accounts: *Newsday,* 2/11/73, p. 4; and *Newsday,* 3/19/76, p. 4. In 1969 when the district was planned, its costs were estimated at almost $300,000,000. By 1974 the costs had more than doubled. Presently, Suffolk County is encountering difficulty in selling bonds to complete the project, even with outside aid.

Chapter 4

THE WEAKNESS OF LOCAL POLITICAL CONTROL

> *Henry F. O'Brien, the Suffolk District Attorney, was charged here today with sexual abuse and misconduct involving a 21-year-old unemployed handyman, in papers filed by the County Police Commissioner, Eugene R. Kelley.*
>
> *Mr. O'Brien immediately termed these charges "utterly false, vicious and depraved." At a news conference here this afternoon, he accused the Commissioner of engaging in a conspiracy to defame me and destroy the credibility and integrity of my office . . ."*
>
> *The Commissioner, who is a Republican, is under investigation by Mr. O'Brien on allegations of "corruption, misconduct in office and other more serious crimes." New York* Times, *Thursday 12/4/75, p. 1.*

Suburban politics have always taken a back seat to the drama of central city political machines and their bosses. Urban analysts have paid some attention to the political process in the outer regions of the metropolis, but not in proportion to their present-day plurality of voters. In the 1950s, for example, observers were struck by the integrated, consensual nature of local suburban government. Its characteristics apparently mirrored the social homogeneity of the upper middle-class white residents. At that time, it was noted that local suburban party leaders led a constituency which shared a wide understanding of the duties and goals of government, and which preferred nonpartisan professional administrations despite heavy Republican enrollment.[1] With increasing population diversity, however, these communally oriented political attitudes began to change. Although analysts in the 1960s still clung to an image of suburbia which conveyed high levels of community participation, some of their observations revealed several features of suburban politics that reflected the relative weakness and limited scope of local political organizations, compared to those of the city and state.[2]

First, it was noted that local suburban politicians did not distinguish themselves and were not visible beyond the borders of their own communities, or, as Scott Greer has said, "They all stand equally high and none are very tall." Second, it was observed that compared to the city, suburban political conflict was not split along party lines. Instead, political issues reflected the outlook Morris Janowitz has called, "the community of limited liability." The concerns of suburban residents seemed restricted to issues that dealt directly with the quality of local community life. The narrow scope of issues, participation, and public administration supported Greer's contention that suburbs were serviced by "toy governments."[3]

Third, the presence of extra-local regional authorities with broad powers, such as HUD or the Port of New York Authority, and county governments which supply local services, such as police protection in Suffolk County, diminished the prestige of local government and reduced its influence in the administration of regional policies and programs.[4]

Many of these more recent observations on the weaknesses of suburban government are valid for Suffolk County and Privatown today. Local government is limited in scope, has low prestige, and is relatively ineffective in meeting the needs of the developing region. The townships on Long Island, for example, have the benefit of autonomous home rule, but they are aggregated in a suburban region which is criss-crossed with a large number of different municipalities and service districts in a confusing mosaic of overlapping jurisdictional lines. This decentralization diminishes the status of local government and often makes neighboring townships compete with each other for scarce tax-producing resources. Administrative fragmentation also makes nonelective regional governmental coordination a task for super-agencies, such as the Urban Development Corporation or the Nassau-Suffolk bi-county planning board, which then have to fight with local interests over home rule in order to plan for suburban development. The weak, fragmented role of local government is also reflected in the low visibility of suburban politics in the mass media, despite the exceptional work of *Newsday,* the daily paper of Nassau and Suffolk counties. Suburban political activities and politicians are not taken very seriously by news media outside the region.

In contrast to these observations, the current image and long-time reputation of Suffolk County and Privatown stress control by a strong Republican party organization. In 1968, for example, the county party was heralded as giving Richard M. Nixon his highest plurality of votes in the United States during his bid for the presidency. The strength of Republican party control, however, may well have been overestimated in the past, and it is presently on the decline as a consequence of suburban development and the growing maturity of the submetropolitan region. Contrary to the Republican image of Suffolk, for example, is the success of Democrat Otis Pike, a prestigious member of the House of Representatives, who recently chaired the House Select Committee on Intelligence. He has been continuously reelected by Suffolk County voters despite his Democratic party affiliation and his somewhat liberal posture.

While Democrats have always been active in local politics, it is true that until recently they were something of a weak sister to the Republicans. This was, however, more a problem of organization than of votes. In the years following World War II, the Democrats in Privatown functioned in an organizational vacuum, even though more and more democratically inclined voters were moving there from the city. During the 1950s, the Republicans retained control of local government by rallying the older residents with appeals to their desire not to be "overrun" by New York City. As late as 1975, Republicans in Privatown could still support a township campaign against a Democratically sponsored reform referendum with the slogan, "Keep New York City out of [Privatown]."

In 1958, however, with the first breath of political scandal in the township, the nascent Democratic party won momentary support from independent voters and took temporary control of local government in the 1960 elections. At this early date, however, they still lacked an effective organizational base, and within two years the Republicans recaptured their control. Finally, in 1973 and again in 1975, the Democrats achieved success at the polls and political control of the township, this time without benefit of large Republican protest voting or any obvious local scandals. Although the most recent elections reflect the effects of both Watergate and some widely publicized local Republican corruption, they are also clear indicators of the emergence of a suburban two-party system. The decline in Republican party strength does not, however, mean that the county is shifting to a Democratically controlled one, or even one in which electoral competition is highly organized and keenly fought. What the shift does mean, is that neither the Republicans nor the Democrats have been able to develop reliable constituencies, and that party control is now weak.[5]

In Privatown the local polity provides little resources in the form of monetary support, competent individuals, or status to the township government. Privatown politicians enjoy minor social privileges. The local parties share an inability to recruit people with higher prestige despite their availability. If anything, it is the Democratic party that attracts individuals of high aspirations who wish to participate in state and national politics. The Republican political organization can best be characterized as having a local "club house" atmosphere. Most of the district councilmen, of whom it consists, are the "chums" or "cronies" of the township's Republican leader, Carl Middleton, a wealthy car dealer, who has controlled the party since 1962. They get together socially as well as on political occasions. There are no bankers, industrialists or highly respected professionals participating in local government despite the presence of such individuals in Privatown. During the time of our case study, 1970 to 1973, the Republican-dominated town board consisted of an accountant, two real estate agents, the director of a funeral home, a local housewife, and an independently wealthy member of the Conservative party, the lone political dissenter. In spite of his role as Republican party boss, Middleton has exercised only weak control over the local political organization. During the past fourteen years, he has engaged in constant in-fighting with various "reform" factions within the party. Following the zoning scandals of 1968, for example, Middleton was censured and called upon to resign by the head of the county Republican party himself. In 1970, Middleton was removed by the county party from his post as chairman of the Suffolk County Water Authority on charges of misuse of confidential information and conflict of interest. Most recently, in 1975, he was indicted by the Suffolk County district attorney on a charge of violating the election law by spending party funds in a primary. Although this indictment was subsequently dropped, factional disputes have continued to produce "dump Middleton" movements.

The weakness of political leadership in Privatown is matched by the limited support of local citizens for suburban government. The general population is issue-oriented and little involved in the affairs of running the township. The local political organization cannot control votes easily by patronage, because township residents, for the most part, earn enough income not to be dependent upon public service for employment. In Privatown patronage itself is scarce. The jobs that are available often go to party committeemen from the 147 election districts or to their wives and relatives. When many people first moved out to the township from New York City, there was an incentive to register Republican in order to obtain summer employment for their children or special favors for their developing communities. This period, however, has now passed. Presently, the Democrats and Republicans seem to share control, and it is no longer clear where the source of township patronage lies.

The limited support of local political organizations can also be seen in voter registration figures. Following the large Republican plurality in the 1968 presidential election, the township registration figures for 1969 revealed that 55 percent of the voters were Republicans, 30 percent Democrats, 11 percent Independents, and 4 percent Conservatives. An independent survey of voter attitudes towards party affiliation, however, carried out in 1968, uncovered the following breakdown in party loyalty: 34 percent Republican, 31 percent Democratic, and 25 percent Independent.[6] The survey reveals that many Republicans expressed apathy towards active political affiliation. They do not appear to be as involved in party support as central city voters and form an unreliable pool of independent ticket-splitters in the polity. Recent trends in voter support can be illustrated by registration figures for the past several years in Suffolk County.

Tables 4.1 and 4.2 show that there is little variation between parties from year to year. Republicans constitute a majority with around 53 percent of the enrollment, 30 percent are Democrats, 3 percent Conservatives; Liberals average close to 1 percent, and Independents around 13 percent. Second, the Privatown figures correlate closely with those for the county. In Table 4.2 we see that Privatown Republicans and Conservatives have slightly more enrollment strength than in Suffolk County as a whole, and, on the average, the Democrat and Independent registration is somewhat smaller. The figures in Table 4.2 are comparable to those of the 1968 voter attitude survey in Privatown.

Table 4.1: Voter Enrollments for Suffolk County, 1969-1976

	Percentage of				
Year	Republicans	Democrats	Conservatives	Liberals	Independents
1969	54	29	3	.7	12
1970	53	29	3	.7	12
1971	54	29	3	.8	12
1972	52	30	4	.9	13
1973	50	29	3	1	14
1974	53	30	3	1	12
1975	54	30	3	1	12
1976	54	30	3	1	12

NOTE: Suffolk County Board of Elections.

Table 4.2: Voter Enrollments for Privatown, 1969-1976

	Percentage of				
Year	Republicans	Democrats	Conservatives	Liberals	Independents
1969	55	30	4	.8	11
1970	53	30	3	.7	11
1971	53	29	3	.7	12
1972	52	29	3	.9	12
1973	52	30	4	1	13
1974	54	29	4	1	11
1975	56	29	4	1	11
1976	56	28	4	1	11

NOTE: Suffolk County Board of Elections.

In order to measure more accurately the extent of voter independence and party support since 1968, Table 4.3 was constructed. It lists voting patterns for the bi-annual Privatown election years 1969 to 1975. Figures are for the local general elections in which township officers, the town board and other officials are picked. Only the data for the town board supervisor race was used, because the elections for other township positions have similar results. Table 4.3 shows voter turnout averages more than 60 percent which is above that of central city areas, which average less than 50 percent in local elections. Second, Republican voter strength is considerably below registration. In each year the Conservative candidate received two to five times more votes than might be expected from township enrollment. The Democrats experienced more variation in support. In one year, 1971, they got less votes than enrollment had indicated, while in the other elections they received more. In 1975, in fact, enough Republicans "crossed over" to elect a Democratic supervisor, defeating the Republican incumbent, who had held that position for almost ten years, and Democrats obtained seats on the town board. In general, these patterns suggest weak party support and considerable ticket splitting.

Table 4.3: Town Board Supervisor Elections

			Percentage Voting		
Year	Total Enrolled	% Turnout	Republican	Democratic	Conservative
1969	82,099	67	47	40	10
1971	77,976	87	50	26	17
1973	112,074	60	48	35	10
1975	100,237	75	40	47	7

NOTE: Suffolk County Board of Elections.

Township Government

Faced with the inevitable increase in population during the previous two decades, the task of the local township political organizations has been to help articulate and meet the needs of the region. In spite of this very critical planning role, the local government of Privatown has pursued the public interest in limited ways without rising to the challenge of rapid growth.

In Privatown we find that the limited success of local government is due largely to the many weaknesses of political control in spite of the considerable advantage of home rule. As a consequence of such weaknesses, party control in Privatown is used primarily to support local government financially and to realize personal gain. This observation raises several questions regarding the ultimate effectiveness of the local government's land-use planning and the continued viability of decentralized home rule in the metropolitan region.

In the following discussion we shall explore the ways in which the operation of local government is utilized as a revenue source for party politics. The zoning record of the town board will be examined in detail in order to discover the interests behind the decisions and the reasons for the inequitable land-use. We shall then discuss how a party leadership that relies on the "spoils system" to support itself financially ultimately subverts its ability to plan adequately for the needs of the expanding metropolitan region. But let us first examine the structure and operation of the Privatown township government.

THE TOWN BOARD

The locally elected government in Privatown consists of a seven-member town board including a supervisor and six councilmen.[7] The members of the board are elected on an at-large basis with three councilmen standing for election every two years. At each of its bi-monthly meetings, the local government makes a variety of decisions and holds discussions on matters related to township business. The board has at its disposal only residual powers of decision-making not covered by the higher levels of government. Its political power arises from its control over zoning, its small supply of patronage positions, and its ability to award contracts for township services, such as sanitation, construction and maintenance of roads, public buildings, and recreational areas. Roughly 70 percent of Privatown's operating budget is raised through real estate taxes which had an assessed valuation in 1969 of 4.2 percent. In that same year, Privatown's total budget was $12,976,527, with 17.7 percent supplied to the township by outside aid from New York State.[8]

At its bi-monthly meeting, the board receives petitions and proposals from local residents and businessmen as well as schedules township meetings to air important issues. Field observations revealed that such meetings were sparsely attended with from thirty to sixty people. Only a cadre of regulars holding diverse perspectives on local government appeared at each meeting. These included representatives from opposition Conservative and Democratic parties, an observer from the League of Women Voters, members of several civic associations, and an occasional observer from the Privatown Housing Coalition. When important issues were discussed, however, especially in connection with land-use, attendance increased greatly and, at times, numbered well over one hundred people. At all meetings public discussion was encouraged by the board and, occasionally, spirited exchanges between councilmen and local residents were observed. In appearance then, Privatown government seems quite democratic as it operates in the light of the open forum of bi-monthly meetings.

THE LOCAL REPUBLICAN PARTY

In spite of this observation, the local government has been overshadowed by the presence of the Republican party leader, Carl Middleton. For the past fourteen years, his organization has exercised control, at times only weakly, over political patronage and the selection of slates for local elections. Except for the recently elected Democratic town board, virtually all the councilmen sitting on the board during Middleton's tenure as township leader have been loyal party functionaires, and they have given his political organization control over at least the board's land-use decisions. The open appearance of town board meetings cannot always mask the operation of this political organization. This was particularly the case whenever controversial land-use decisions were discussed, and open challenged to the political leadership were made by housing coalition groups and local civic associations.[9]

Lacking much patronage and support from local citizens or the larger business community, the party has limited financial resources. The task of making local government work, which falls to the township political organization, means making the best of the political resources that are available. In the case of Privatown, control over the decision-making apparatus of local government and, especially, its zoning and land-use regulatory powers has become the major revenue source for the party. The political process is highly dependent upon the construction industry, which needs local government cooperation for permits, inspections, and public decisions on zoning in order to commence work. Town boss Middleton and his party exploit this need in a business-like way to generate money for party support and personal gain.

Local party control of the town board's decision-making powers has operated in Privatown under the following constraints. On the one hand, the party has to appear to be responsive to the public trust and the opinions of voters, because it can seldom manipulate resident sentiments. This is important to Middleton, who exercises only weak party control and continually fights with factions wishing to depose him. Party dissidents as well as opposition candidates are quick to use adverse public opinion against the political leadership in elections. On the other hand, the local party organization must also be responsive to the needs of the business community. The local Republican party does not represent the elite economic interest in the township. Most of the businesses are operated by national corporations and banks, such as the owners of the large shopping malls, giant department stores and the developers of some of the township's housing. The heads of these corporations come and go in Privatown and conduct their business with little regard for local party bosses or political machines. While entrepreneurs, such as the locally based housing developers and real estate speculators, do have some evident connections with the political organization, this connection is absent when we consider the corporations and banks which operate at the national and international level. Consequently, township government must be responsive to the interests of both public and party sentiments as well as to business needs. The local political organization has developed a decision-making strategy which acknowledges these constraints and which exploits control over land-use to support itself. This view of political decision-making and the operation of local government differs from traditional elitist and pluralist conceptions of community power. Our view of local government concentrates on its function as a separate sphere of social organization and focuses on the weakness of present-day political parties.[10]

The Problems of Political Control

The zoning process in Privatown has the appearance of being an open and democratic one. Every land-use decision involves businessmen, who make most of the proposals, and the participation of local civic associations and private citizens, who are allowed to air their views regarding the project's impact. In addition, the town board also receives an evaluation of a petition from the township's planning board and, in some cases, an analysis from the professional planners of the regional planning board as well. Each decision, therefore, involves political, economic, social and technical planning considerations. It is left up to the town board to weigh such factors and make the decision. Although professional planners may be called upon to explain their visions of the future, it is left up to the town board to *implement* land-use decisions and to respond to the many kinds of influence in the township. It might be argued that this separation of function between planners, citizens, and political implementors is as it should be, because most of the affairs of government are conducted precisely in this manner. As we shall see, however, this exercise of local power through representative democracy results in unforeseen and often unpopular consequences.

I have examined 372 zoning cases covering the years 1968-1971, a period with rapid suburban growth. With the aid of local informants, interviews of residents and local newspaper reporters in a two year field study, information was obtained on the community interest groups participating in these decisions and the benefits they received. In considering the re-zoning record for the four year period, the initial research question was to what influences was the board most responsive. In the majority of the 372 cases (over 60 percent) the board acted in agreement with the recommendation of its own planning board and with the desires of the local residents. Whenever these desires were the same as the interest of the petitioner, there was also complete concordance in the decision-making process. This often happened. In approximately 30 percent of the cases, however, the town board rejected the recommendation of its own planning board. As Table 4.4 shows, this pattern did not vary greatly from year to year. In these cases the technical considerations were disregarded, and other influences were more important. In appearance, then, a pluralist

Table 4.4: Proportion of Zoning Cases in Which Town Board Decisions Affirmed Local Planning Board Recommendations

Years		Agreement	
1968		74%	(98)
1969		67%	(71)
1970		67%	(92)
1971		69%	(111)
	Total . . .		(372)

NOTE: Gottdiener, **Social Planning and Suburban Development.** See also note 11.

pattern is suggested by the bulk of land-use decision-making. Often, however, personal interests are served at the same time as the public trust is fulfilled. In order to identify other factors of influence which might be at work, the cases of disagreement were examined in detail, while cases of complete concordance were temporarily ignored.[11]

A check of the public hearing records reveals that in 60 of the 372 cases where disagreement existed, the local citizens groups indicated some preference. In the remaining cases *no* civic association went on record with an opinion. We are left, then, with this minority (16 percent) of the total number of zoning decisions.

In Table 4.5 we have summarized the re-zoning records for the sixty cases. The data reveal that in a slight majority of these (thirty-two), local civic associations supported the planning board *against* the final decision of the town board. In these thirty-two cases, the board's action to overrule the planning board seemed to disregard the major sentiments expressed in its decision-making procedure, i.e., the town board went against *both* its local planning board and the desires of the polity as represented by local civic associations.

In the absence of more detailed information we can only infer that in the remaining twenty-eight cases the social input of civic associations was given more weight by the board than the technical recommendations of its planners.

Cases in which public officials ignored the wishes of civic associations and planners were the most promising for obtaining a picture of the patterns of influence. As it turns out, these thirty-two cases are essential to

Table 4.5: Breakdown of Concordance Between Town Board and Local Civic Associations for Zoning Cases in Which the Town Board and the Local Planning Board Disagreed, 1968-1971

	Percentage of Cases in Which Town Board and Planning Board			
Year	Agreed	Disagreed	Ambiguous	N
1968[a]	62	38	0	8
1969	20	60	20	10
1970	25	60	15	20
1971	14	72	14	22

NOTE: See Table 4.4.

a. 1968 was the only year in which the town board supported the view of citizens against the planning board most of the time. It is significant that, as a result of political scandals during that year, in which a number of town officials were under indictment, the board was under close scrutiny.

the township for planning purposes, as well as critical to its ability to control submetropolitan development, and they are the most controversial ones arising during the period studied.

Out of the thirty-two discordant zoning cases only three did not represent a change of land-use category, i.e., twenty-nine of them altered the zoning pattern. Of the thirty-two discordant cases, nineteen were changes in classification for parcels to recipients who were identified through fieldwork investigations as either being party supporters or direct business associates of town boss Middleton. The largest category represented changes of parcels located in residential single-family or limited commercial zones to multiple-family dwelling classification. This is the most lucrative re-zoning of land because the township has a severe shortage of apartments brought about by the widespread opposition to such housing in the region. Eight of the cases, or 25 percent, were changes to the multiple-family category. Of these eight, seven were received by active business partners of Middleton.[12]

Eight other cases involved changes from residential to limited commercial categories along main residential roads. Commercial re-zonings in residential areas create strip zoning of residential roads and often pave the way for wholesale changes of zone to commercial purposes. Of these eight re-zonings, three were received by business associates of Middleton and by the township boss himself. The remainder were all obtained by Republican party supporters. Although thirteen of the re-zonings, or 40 percent, went to corporations or individuals with *no* known connections to the local political organization, this does not necessarily imply that the party did not receive compensation for the town board decision. As we shall see, payment to the party can take a number of forms.

The business associations of the political leadership and the patterns of influence identified in the analysis of the discordant cases were then used as a basis for the analysis of all the re-zoning cases which occurred during the four year period. This work was supplemented by field investigations. Local informants indicated the names behind most of the development corporations operating in Privatown and their connections with the political organization. Interviews with newspaper reporters, lawyers, builders and real estate agents helped to complete our view of the patterns of association. This investigation indicates that the town boss Middleton and a select group of associates utilize political control as a business resource to benefit directly from rapid suburban development.[13] For example, in 1968 *Newsday,* the Long Island newspaper revealed that Middleton, working closely with his friends, obtained over $1,000,000 in a series of township re-zonings. His connections worked through partnership in at least twelve different corporations active in real estate development. In addition to this personal gain, political control of the zoning power is also utilized to generate revenue for the party from grateful businessmen who have obtained favorable decisions. This holds for many cases in which the planning board and residents agreed with the town board decision. There is, then, a network of connections between party officials and private businessmen, creating many opportunities and conduits for financial transactions which enable the local political organization to appropriate value in land-use decisions for personal and party gain. It is not simply corruption, but also a way for the relatively weak local party and its political leadership to support themselves financially.

The following relationships are suggested by the data, although complete proof cannot be given:

1. Lawyers with political connections to the leadership, as identified by informants, appeared to be more successful in obtaining re-zonings than lawyers without traceable connections. From observing the more informal aspects of decision-making it appears that a small group of lawyers tended to be brokers of influence between political leaders and those wishing to obtain favorable land-use decisions. At times, this small group of lawyers might also join politicians and speculators as the third party in a business venture, realizing gain from a re-zoning. Of the fifteen lawyers who handled the cases of discordance, six were business associates of Middleton. In addition, fees to key go-betweens in the town were also used as conduits for contributions to the local party organization.

2. Another indication of these relationships is given in Table 4.6. The performance record of lawyers in *all* the successful re-zoning cases is summarized for the four year period.

Table 4.6: Performance of Lawyers With and Without Political Connections in 169 Successful Re-Zonings, 1969-1971

	Number	Number of Lawyers	Average Number of Successful Cases by Lawyer
Successful Cases by Lawyer With Connections	72	19	3.8
Successful Cases by Lawyer Without Connections	97	49	2.0
Total Number	169		

NOTE: See Table 4.4.

Obviously, those lawyers identified as having definite connections with the political organization have a higher average number of successful cases than those without similar connections. Although comprising less than 30 percent of the total number of lawyers, those with connections handled almost 50 percent of all successful re-zonings. This relationship is strengthened by the knowledge that many of the successful attorneys are business associates of Middleton and by the fact that the township has other successful attorneys who do not handle re-zoning cases at all.

3. Informants indicated that some relationship usually existed between petitioners for changes of zone and a member of the political organization. Those associated with the leadership worked through connections to obtain favorable decisions. Many businesses, however, received re-zoning without such influence. Successful petitioners can, therefore, be divided between those with and those without traceable ties to the political leadership.

Efforts to find out the exact connections between petitioners and the political leadership encountered great difficulty. Nevertheless, identification of such a relationship was possible in 305 out of the total 372 cases. The breakdown of these Privatown cases is given in Table 4.7. Those *with* connections were successful in 63 percent of a total of 115 proposals. Those lacking known business or political ties were successful in 51 percent out of a total of 190 applications for change of zone.

These data do not necessarily prove that political connections were instrumental in obtaining re-zonings. Although fieldwork informants have indicated that such a direct relationship *epitomized* the operation of local political control, the interest of civic groups often coincided with those of people with connections. It is precisely for this reason, that we must return to our study of the thirty-two cases of discordance where the town board went against the public demands. The actual number of cases in which personal or political gain is the major interest behind local political

Table 4.7: Comparison of Success of Petitioners With and Without Connections, 1968-1971, 305 Town Board Cases

	Total Cases	Successful	Unsuccessful
With Business or Political Connections	115	63% (73)	37% (42)
Without Business or Political Connections	190	51% (98)	49% (92)

NOTE: See Table 4.4.

decision-making may be a quite small and seemingly insignificant fraction of the total number of cases. What is important for the purposes of our discussion is to point out, first, the ability of the local political organization to gain from these decisions and, second, that in the case of Privatown the power of local government to control land-use is the major source of revenue for local politics.[14]

A qualitative study of the thirty-two cases reveals a number of key effects in the pattern of growth for the township. These cases involve widely recognized violations of existing land-use categories and result in distinct inequities. These include, for example, the construction of multiple-family dwellings, shopping centers, and gas stations on legally limited-access highways; office and commercial buildings in residential areas; and the commercial re-zoning of public land. Over time, the small number of these decisions aggregate and help define a land-use pattern, which represents unplanned growth or suburban sprawl, and which operates to undermine further citizen and planner attempts at controlling the process of suburban development. The impact of such decisions can be illustrated, for example, by the commercial strip zoning and residential development of Nesconset Highway. This four-lane road was originally proposed and approved by the voters as a nondeveloped limited-access highway in the 1950s. It was considered a much needed solution to the traffic congestion plaguing the developed strip-zoned highway 25, which runs across the northern portion of Suffolk County. Nesconset Highway was proposed as a by-pass for north-shore traffic and as a connection with the terminus of the Northern State Highway in Smithtown. Shortly after construction began, however, the noncommercial thoroughfare was invaded by speculators and re-zoned for business and residential use by the Privatown town board at numerous locations along its length, and several giant shopping malls have been built. Many of these re-zonings were protested vehemently by local civic associations. Others, however, were passed without much resident resentment. In the latter cases, few people were actually living in the adjacent areas concerned because the development of housing itself was limited until the new highway was built. After its construction, speculators cashed in on their holdings. Developers, such as Levitt and Sons, built large housing communities and the thoroughfare became an artery for commercial and residential traffic in the familiar pattern already existing on highway 25. Presently, Nesconset Highway is a congested, commercial road with heavy commuter traffic to new residential development, and it is the scene of numerous accidents. It has taken on a shape and pattern of use which would exist, had there been absolutely no planning or town board zoning decisions at all.[15]

Local Government as Corporate Business

Due to the weakness of political control and to the opportunities afforded by rapid suburban growth, local government is highly receptive to the interests of the real estate and construction industry. The Privatown Republican party has adopted a strategy of decision-making which uses zoning powers for financial support, while recognizing the limitations of local government control. These limitations arise from public awareness of corruption, minority factions within the party, growing electoral competition from the opposition, and the broader activities of national corporations and banks. The opportunities for gain are carefully weighed against the political risks in the private caucuses of loyal Republican councilmen held prior to town board meetings at the direction of boss Middleton. Such occasions give the political leadership the opportunity to discuss how votes should be cast in re-zonings so that political costs and benefits can be assessed. Like other features of community political control, this procedure is quite legitimate and can be defended under our system as a party caucus of the political leadership. These meetings, however, have enabled the councilmen who are loyal to the local political organization to act in the best interests of the party as well as the polity, and to plan a decision-making strategy which would not jeopardize their political future. Because Middleton's "cronies" have often enjoyed a voting majority on the town board, such meetings also enabled them to decide how each particular vote should be cast. In this way, specific councilmen up for reelection might dissent from an unpopular decision and give the appearance of independent voting without endangering party control and with the blessing of the political leadership.

Party control of the town board has been exploited for two distinct purposes. On the one hand, the granting of favorable zoning decisions and construction permits can be reciprocated by businessmen in the form of payment to the local leadership and the party. This has most often been accomplished through political contributions. Large-scale developers of housing, national corporations and banks, as well as commercial interests in shopping center malls generally have used this method of compensation for favorable political decisions. Often payment has been effected through the purchase of blocks of tickets to local party functions in addition to giving campaign contributions. This impersonal type of payment has been utilized because of the relative lack of direct connection the large corporations have to the local area. A personal appearance by the representative of one of these national enterprises at a local party function would be an extreme rarity. Furthermore, the ability to exercise this power needs only to be demonstrated occasionally to the business community for them to become regular supporters of the local political organization. The latter is less in the business of selling specific votes to national enterprises than in offering an attitude of cooperation and sensitivity to profit-making interests in exchange for financial support to the party. Consequently, yearly political contributions have been made to insure the continuance of such sensitivity over the tenure of office with the implication that town board receptiveness could always be withdrawn or cooperation made more difficult in the event that business ceased its support. Just as is the case with the operation of social control through other bureaucratic and legal regulations, local political control can operate effectively by structuring activities through its *implied* powers, even when they are not directly used.[16] In these dealings, the zoning power of the town board and the political regulation of the construction industry have been exploited to enrich the party's campaign fund. This enhanced its competitive position at the polls and helped it to remain in power.[17]

Control over the decision-making appartus of local government enabled the holders of office to perform a second function. A persistent fraction of all town board decisions directly benefited the economic position of a select group of local political leaders and their business associates. The people involved in the use of public office for personal gain did not represent the business elite in the township. They were a small number of speculators, lawyers, politicians, investors of capital, and local housing developers. The close relationship between Middleton and this group was not a "conspiracy" of the social elite, but, rather a collection of entrepreneurs who used local government control to make money the way businessmen use economic organizations. Other political and economic interests operating in the township and region bargained with this select group whenever the necessity arose. Economic and political exchange characterized the interaction between these individuals and not a confluence of interests.

Personal gain from political control of land-use and the presence of rapid suburban development could be realized in many ways through numerous financial avenues. At times, money was made by direct re-zonings involving enterprises in which some of the members of the select group were partners. A controversial re-zoning for multiple-family housing, for example, went to several of Middleton's business associates. The value of the land could also be manipulated through local tax assessment procedures or the proposal of township government developments so as to directly benefit the holdings of this group. In one instance, a parcel of land owned by the leadership's associates was purchased by the township for recreational use at many times its market value. Finally, contracts for construction and other services, such as legal representation, were often given to this select group by corporations in return for favorable consideration of re-zoning proposals. In Chapter 5 we shall examine several businesses connected with the turnover of land, such as real estate and development firms, which were associated with the political leadership and which capitalized on opportunities for personal gain from land-use decisions.

As reported many times in *Newsday,* the opportunities for making money from land have been created continuously over the years by the growth patterns of rapid suburban development and have been seized by this select group in numerous deals that have yielded considerable profits.[18] In fact, conflict of interest, as it has often been called, seems to be characteristic of a sizable percentage of all zoning decisions. Furthermore, except in years of scandal, the political leadership has appeared able to take considerable risk in conflict of interest during the past two decades of suburban growth. The recent elections in 1973 and 1975, which were won by Democrats, however, do show the increasing discontent with the Republican record in Privatown. Nonetheless, as of this writing, the political leadership of the party is still intact.

The term "corruption," which has been used to describe conflict of interest, and in most cases to dismiss its significance, seems too pristine for what was apparently an important and *inherent* feature of the exercise of political power in Privatown.[19] Suffolk County as a whole has had a long history of political corruption. Middleton and his associates were but one of several groups that established this rather long tradition of the "spoils system" and the operation of political control. What was perceived by the public and the press as periodic scandals, was in reality an institutional feature of everyday political life, and no party or individual necessarily deserves to be singled out. Most significant for our study is the fact that this tradition of corruption in county home rule arose from the political control over land-use coincident with the period of suburbanization and the need of weak political parties to use zoning for financial support. Thus, in 1955, countywide scandals arose out of a land-grab scheme in which eighteen indictments were handed down, and at least fifteen officials resigned. Those indicted were accused of "illegal land purchases and speculation, illegal re-zonings and the participation in profit sharing kickbacks to other county officials."[20] In 1958, several years *before* the present Republican leadership arrived, Privatown's supervisor and two other township officials were indicted and subsequently convicted on charges of illegal zoning decisions and bribery. The officials were also found guilty of running a zoning "racket" in which they sold lucrative down-zonings to interested business associates. Furthermore, the conditions under which this racket arose were precisely the result of the response to the initial influx of population. The down-zonings occurred after the school districts had pressed for and won the first in a series of townshipwide up-zonings of residential land, which they hoped would control rapid growth. They were discussed more fully in Chapter 2. Once this occurred, down-zonings by the board to more intensive business and higher-density residential uses became highly profitable.[21]

The use of political decision-making for financial benefit in Privatown has several other characteristics: (1) Unlike the traditional political machine, local suburban government was little involved in electoral politics on a mass basis.[22] Because political life in Privatown was a dead end, the leadership was more concerned with using public decision-making to support the party. (2) The township boss and his associates did *not* attempt to squeeze out every cent possible from their control of land-use[23] because the political risks were much too high. There was also a limit on the tolerance of national businesses. If the price of cooperation was set too high by local leaders, the corporations could always locate elsewhere in the county. Due to the critical nature of land-use decisions, however, it was necessary to exploit control in only a small fraction of decisions in order to realize significant monetary gain. One reason for this is, as we have already indicated, that the implied use of such powers can be as effective a fund raiser as their actual application. Strict control over decision-making, therefore, was *not* necessary in order to make money for the local political leadership. This is a major reason why the overwhelming majority of town board decisions had the appearance of siding with public opinion. (3) The operations of the local political organization are as diverse as the activities of any corporation in the pursuit of gain. The web of connections characterizing the activities of Middleton and the select group of lawyers, speculators, and developers surrounding him made use of both public and private resources. That is, businessmen used local government to create favorable opportunities, to supply them with benefits from public works, such as highways, and to supply them with information on the housing and land-use needs of the community, such as the patterns of township growth. At the very same time, the political organization used the capital and activities of private business. There appears, then, to be little difference between the actions of public and private individuals

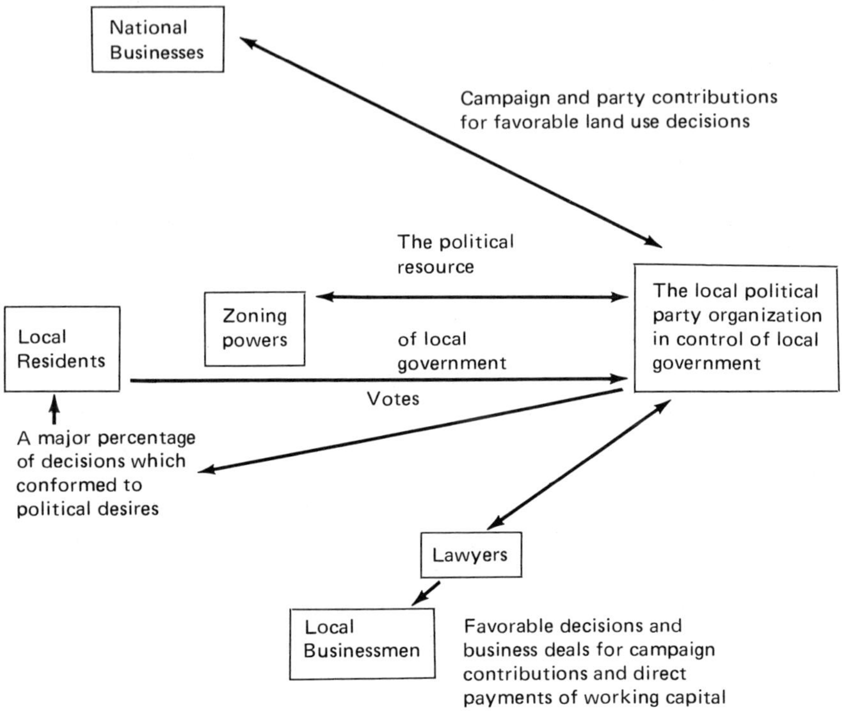

Figure 3: POLITICAL RELATIONSHIPS IN PRIVATOWN

at the local level. The relationships between the various groups involved and the exchanges which characterized their interaction can be illustrated by Figure 3.

The structural relationships indicated in Figure 3 can be summarized as follows:

Local residents:

Got: zoning decisions. A major proportion of these conformed to their desires about the growth patterns of the township, e.g., keeping blacks out by not voting for low-income housing or voting for up-zonings of residential land.

Gave: votes to the political organization in power.

Local businessmen:

Got: favorable decisions and inside information on the programs and policies of government and the national businesses in the township, which then enabled them to cash in on the opportunities that are created.

Gave: payoffs to the party and, in some cases, business partnerships with select individuals in the political organization.

Local Republican party:

Got: campaign contributions.

Gave: favorable decisions through control of local government within the limits defined by the constraint operating on decision-making.

National businesses:

Got: favorable decisions from a local political organization to which they do not have or need to have any direct ties.

Gave: campaign contributions.

Conclusion

In Privatown the stereotype of strong Republican party control, a local suburban government supported by a broad consensus of community interests and the active participation of residents seems to be a myth. The local political organization must rely on scarce resources, limited government powers, and the absence of much patronage or party loyalty. The growing number of voters with independent party affiliation and the wide practice of ticket-splitting indicate that Privatown political parties can no longer depend upon reliable constituencies. In addition, the local political organization can no longer rely upon the public to support the financing of campaigns. Nonetheless, campaigns depend very much upon ample funding, because they are carried out largely by the mass media. In contrast to the ward system of cities, in Privatown there is little organizational power at the local level. In the place of local precinct captains who contact the residents of local neighborhoods, suburban political organizations must rely on mass mailings, newspaper advertisements, and spot radio and TV announcements. In the New York City area this is quite expensive.

As a consequence of these factors, as our study showed, the local political organization took advantage of rapid submetropolitan growth and was heavily dependent upon real estate and construction interests as a means of party support. The power of zoning and the ability to regulate home construction exercised by the party in control of government became the major revenue sources for local politics. Local government control over land-use, furthermore, was used to create value in real estate. This ability was exploited in a strategic way as a very direct means of acquiring wealth for political leaders. The use of political control for personal gain strongly offends the voters whenever it is discovered, yet, the temptations of doing so presented by suburban development were apparently too enticing to be turned down. This further undermined the already low prestige of township politicians and weakened party control.

One important effect of the limitations of local government is that it subverts the otherwise desirable feature of home rule, i.e., local political control of the planning function. The decisions made by the political organization in pursuit of gain have accumulated over the years to contribute to the social problems of the region. The land-use pattern of spread city, discussed earlier, can be viewed as the aggregation over time of many politically expedient zoning decisions. The thirty-two cases in which business interests prevailed over planner and resident recommendations illustrate the most obvious violations of sound planning principles. The need of the local party to cater to the short-sighted desires of the polity regarding housing restrictions and their strong demand for up-zonings in order to limit the rapid residential development of the region must also be included in the list of expedients. Exclusionary zoning and the racial and income inequities reinforced by the spread city pattern have become evident consequences of this legacy. Local government decision-making combined short-sighted vision with a subversion of the regional planning process. There was, thus, an absence of any viable future orientation to guide political decision-making in Privatown, and land-use took on a pattern which gave the appearance that no planning occurred at all.

Political leadership with a limited business perspective on local government and in control of weak party organizations inevitably fails to meet the needs of a rapidly growing region. Furthermore, local political control and the widely heralded virtues of decentralized "home rule" can hardly be viewed as the appropriate mechanisms through which suburban growth problems can be solved. Yet, recent government policies, such as revenue sharing, and voter sentiments in favor of local community control seem to constitute strong support for the observed state of affairs.

A second observation of our case study concerns the nature and general function of local political control. In Privatown the political organization did not seem to represent either the combined interests of the general public or the specific interests of local businessmen. Local government control, therefore, could not be explained by either pluralist or elitist arguments. Rather, there was a good deal of autonomy attributable to party control of government, and the political leadership used the degree of freedom afforded by representative democracy for party and personal gain. Thus, the local political organization emerged as a social organization which had its own interests, quite aside from business or public sentiments. It used the resources of local government decision-making and the control over land-use to create values of its own in addition to servicing citizen and business needs. This is neither a vulgar Marxian nor pluralist conclusion. It says that the local political organization can attain a limited autonomy by controlling the apparatus of government, and that it may use such control for its own purposes.[24]

As a consequence of moderation in the exercise of political control for personal and party gain, substantive questions on the nature of our system of local government needed never arise. Party problems occurred only when the tolerance level for political corruption was reached by the voters. In between the periodic outbreaks of scandal, most residents were content with the feeling that government works well for them. Scandals, therefore, merely served to increase political alienation and foster a weak form of electoral competition between parties without creating public demands for reforms.

NOTES

1. The most extensive discussion of this view can be found in Robert C. Wood, *Suburbia, Its People and Their Politics,* Boston: Houghton Mifflin, 1959. See also, William H. Whyte, Jr., *The Organization Man,* Garden City, N.Y.: Doubleday Anchor, 1956; and Edward C. Banfield and James Q. Wilson, *City Politics,* New York: Vintage, 1963, pp. 240-242. The city-manager form of government which stressed nonpartisan, communal values was conceived of as the suburban prototype of the 1950s.

2. See, Scott Greer, "The Social Structure and Political Process of Suburbia," *American Sociological Review,* 25 (August 1960), pp. 514-526; and *Governing the*

Metropolis, New York: John Wiley, 1962; William A. Dobriner, (ed.), *The Suburban Community,* New York: Columbia Univ. Press, 1969; Charles Gilbert, *Governing the Suburbs,* Bloomington: Indiana Univ. Press, 1967.

3. An application of Janowitz's concept to the suburbs is found in Greer 1970, op. cit. The original term is used in Morris Janowitz, *The Community Press in the Urban Setting,* Chicago: Univ. of Chicago Press, 1952, especially Chapter VII.

4. See, Robert C. Wood, *1400 Governments,* Cambridge, Mass.: Harvard Univ. Press, 1961, for a discussion of the New York City metropolitan region. For a general discussion of governments in the emerging metropolitan regions of the United States, cf., Alan K. Campbell and Judith Dollenmayer, "Governance in a Metropolitan Society"; and Joseph F. Zimmerman, "The Patchwork Approach: Adaptive Responses to Increasing Urbanization," in Amos Hawley et al., *Metropolitan America in Contemporary Perspective,* New York: Halsted Press (a Sage Publications book), 1975.

5. Polity support is weak in two distinct ways: First, there are fewer voters turning out in local elections, cf., Pranay Gupte, "Voter Apathy of Concern in Suffolk," *New York Times Sunday, BQLI,* 11/2/75, pp. 1, 9. Second, voters are registering or labeling themselves as independents and dropping active party support. These two trends appear to be occurring nationally as well. They are discussed in the context of Suffolk County and Privatown more fully below. There is evidence, however, that the Nassau County G.O.P. still remains strong, but Nassau has a strong Democratic party as well, and it is difficult to make comparisons with Suffolk.

6. James R. Hudson, "Surveying Community Attitudes," in Dieter K. Zschock, (ed.), *Brookhaven in Transition: Studies in Town Planning Issues,* Stony Brook, N.Y.: SUNY at Stony Brook, 1968, p. 50.

7. The town clerk, receiver of taxes, and superintendent of highways are also elected members of township government. In addition, the town board controls 20 municipal service commissioner jobs that are appointed, including parks, planning, sanitation, recreation, and tax assessment.

8. *Your Town Brookhaven,* New York: League of Women Voters publication, 1969.

9. See, for example, the Oak Forest case study discussed in Chapter 6.

10. It is not our intention to contribute to the community power debate. There are ample surveys of the literature in the field, cf., Willis D. Hawley and Frederick M. Wirt, *The Search for Community Power,* Englewood Cliffs, N.J.: Prentice-Hall, 1968. Our view of government is a structural one which considers the power of political decision-making itself as an autonomous productive resource. This perspective has been discussed in several studies, albeit in a limited way. See, for example, Anthony Downs, *An Economic Theory of Democracy,* New York: Harper & Row, 1957; Norman Frohlich, Joe Oppenheimer, and Oran Young, *Political Leadership and Collective Goods,* Princeton, N.J.: Princeton Univ. Press, 1971; and Henri Lefebvre, *The Explosion: Marxism and the French Upheaval,* New York: Monthly Review Press, 1969.

The studies of both Downs and Frohlich et al. are deductive analyses. The former deals with financing party support, but does not consider conflict of interest and is especially limited by the failure to consider how corruption can be used for party support. The latter is a weak analysis because it only deals with leaders and neglects the social web surrounding any party organization, i.e., "the real world." Finally, Lefebvre's case study of France only briefly suggests the important issue involved:

the function of the state in modern democratic society, but does so in a promising way.

11. These data were obtained from the minutes of the town board meetings. This material along with field work on political influence is analysed in closer detail in M. Gottdiener, *Social Planning and Suburban Development,* Ph.D. dissertation, SUNY at Stony Brook, August, 1973.

12. Ibid.

13. Ibid.

14. Several other studies have indicated the role of economic considerations in political zoning decisions. See, for example, Otto Davis, "Economic Elements in Municipal Zoning Decisions," *Land Economics,* vol. 39, pp. 375 ff, (1963), and Davis et al., "Urban Property Markets," *Journal of Law and Economics,* vol. 10, 79 (1967); and the review article Franklin J. James, Jr., and Oliver Duane Windsor, "Fiscal Zoning Fiscal Reform, and Exclusionary Land Use Controls," *AIP Journal,* (April 1976), pp. 130-141. Most of this work, however, does not address the critical issue of the state's ability to support itself through decision-making, but concentrates, instead, on fiscal zoning, land value, and local development problems, i.e., economic concerns.

15. Privatown's poor planning record may not by typical of other local government uses of zoning powers. There are several general discussions of the use of political control of land-use for planning. See, for example, Charles Haar, *Land Use Planning,* Boston: Little, Brown, 1959; and "Zoning for Minimum Standards: The Wayne Township Case," *Harvard Law Review,* vol. 66 (1953); Norman Williams, "Planning Law and Democratic Living," 20, *Law and Contemporary Problems,* 317 (1955). For an earlier study of the same region, see, Herbert Scarrow, "Political Control of Zoning Decisions," in Zschock, op. cit., which also indicates the undesirable effects of town board planning. There is also some evidence that Suffolk County itself has failed to plan for its own government expansion and has constructed buildings that have helped mindless sprawl patterns, cf., *Newsday,* 4/25/73, p. 5.

16. This function of bureaucratic rules as possessing implied powers of social control is discussed in Alvin Gouldner, *Patterns of Industrial Bureaucracy,* New York: Free Press, 1954, pp. 173-174.

17. Downs, op. cit.

18. In 1968 *Newsday* ran a series of articles on these deals, cf., 5/1/68, 5/2/68, 5/4/68, 5/6/68, 6/24/68, 8/7/68, and 7/8/69. This information was obtained from *Newsday* morgue clippings, and page numbers were not indicated.

19. For a general discussion of political corruption, see, John A. Gardiner and David J. Olson, *Theft of the City,* Bloomington: Indiana Univ. Press, 1974; and John A. Gardiner, *The Politics of Corruption,* New York: Russel Sage, 1970. The former has extensive references and a bibliography on the literature in this area. Much of this work ignores the suburbs.

20. *Newsday,* 5/3/68, p. 3.

21. Information of the land scandals of the 1950s was obtained from *Newsday* clippings covering a period from 1955 to 1960. This public ritual was acted out again in 1968 and 1975.

22. Unlike Mayor Daley, for example, Middleton does *not* run for a township office. His political base derives from being elected as a Republican party committeeman from a local district.

23. The same attitude of aiming for a reasonable level of profits, or "satisficing,"

and not maximizing, is also a characteristic of large corporations. See, Herbert Simon and James March, *Organizations,* New York: John Wiley, 1967, pp. 140-141.

24. Local government has the ability to be responsive to society's business and citizen needs, but also to separate from society and the role of public servant, in order to create values of its own. Party control does not "dominate" political and economic resources in the town, because these interests are the sources of its support. However, it also does not arbitrate negotiations or share the benefits of growth in an equitable fashion as the pluralists would argue. This dual aspect of government has important consequences for viable planning in both democratic and socialist societies. The planning bureaucrats, or government decision-makers in communist countries, for example, have been considered to be the new ruling class, cf., Milovan Djilas, *The New Class: An Analysis of the Communist System,* New York: Praeger, 1957. See also Lefebvre, op. cit., for the French case.

Chapter 5

SUBMETROPOLITAN PLANNING:

Developers, Speculators, Investors, and Banks

In our society control over land-use is largely the responsibility of local government. As we have seen, however, political control of the zoning power leads to social problems even when politicians are fairly responsive to the short-run wishes of suburban residents. Concentrations of poverty pockets, commercial strip-zoning, traffic-congestion, and waste-disposal problems are by now endemic in most suburban areas. The local government's limited ability to avoid such inequities is compounded by the fact that many important planning decisions in Suffolk County are made directly by private businessmen. As a consequence, suburban growth is often viewed by analysts as chaotic and unplanned, because private activities appear disconnected from political issues or regional policies. This is one of the many misconceptions fostered by early studies of spread city. Metropolitan expansion actually involves large-scale planning and resource management by private builders, real estate developers and banks. These planned actions are tailored to the various regulations of the extensive planning apparatus existing at all levels of government, and are also supported by federal programs and subsidies, such as FHA and HUD housing programs.[1] As Leonard Downie, Jr., recently observed:

> Suburban development has not been happenstance. In fact it has been exactly as real estate speculators, builders, bankers and even suburban home buyers have wanted, because it has been profitable for them. They have forced government at all levels to plan suburban growth their way through the administration of zoning laws, construction of highways, laying of sewers, writing of tax laws, and supervision and subsidy of mortgage banking.[2]

While we do not necessarily agree that influence operated only upon the government and not from it, what is of direct interest in the study of submetropolitan growth is the way in which both private and public planning efforts have melded to create the contemporary landscape of a seemingly unplanned environment.

Developers of Housing

The planning decisions on land-use during the suburban transformation were political, but the actual altering of the physical environment of Nassau and Suffolk counties was carried out by the housing developers of the private sector. There is no doubt that homes preceded most industry in the earliest stages of Long Island's suburban growth. It was at this point that suburbia came to signify "bedroom communities" dotting the countryside. Later, the massive flow of people, commerce, and industry to Suffolk County created a different phase of development. Suburbia became a submetropolitan region encompassing a complex range of social activities and community types. The regional land-use patterns presently observed were carved out by speculators and developers who altered the presuburban rural acreage on an immense scale. In fact, as early as 1940, enough undeveloped lots around New York City had already been platted by real estate interests to accommodate the entire New York City population of over seven million people.[3]

Regional construction of such magnitude, however, proceeded within existing zoning regulations. The constraints of land-use decisions and building codes were the given conditions to developers. They could build only according to a variety of zoning patterns and had to tailor their plans to suit the local construction regulations. In Suffolk County such restrictions were well defined and administered by local government, although they were not comprehensive and did not provide for sewers. Thus, to the skeleton of political planning decisions, the builders, developers, and speculators added the stock of housing. By the 1960s the result was being called suburban sprawl.

Most of the development of Long Island has proceeded through the construction efforts of locally-based firms. In Privatown, during World War II, building primarily occurred on single plots of land owned by private individuals. The outcome was "custom-built" homes. A notable exception to this was in the village of Patchogue, where some apartments were constructed to house the workers in the village's factories. Single-family homes were built in patches, mainly by custom builders, and each community was separate from the others. Inaccessibility and lack of roads plus difficulty in the supply of construction materials made it financially prohibitive for anything but small developments to occur.

A qualitative change in the mode of house construction took place after the war and was pioneered on Long Island. This change involved the large "development," in which a builder would move into a previously undeveloped area and build a substantial number of houses, and, in addition, supply the infrastructure of roads, utility lines, sidewalks, and sewage disposals, such as in the communities built by Levitt and Sons. From that time on, the landscape of Long Island was destined to appear in the now accepted pattern of "spread city," which is directly attributable to the change in mode of house construction.

The concept of building moderately priced housing on an extensive scale was utilized as early as the 1920s. In 1924, the first of such communities was built by the City Housing Corporation (a limited dividend company) in Queens, at the time an undeveloped area. It was called Sunnyside Gardens and along with its architect, Clarence Stein, it became famous as the first U.S. community built according to modern planning principles. Also for the first time, the developers supplied the infrastructure (now called the development-site requirements) for the houses.

After the war, the housing market was opened to an entire generation of consumers. Returning veterans were benefited by government-sponsored housing loans at very low rates of interest. In the center of Nassau County, a few miles east of Sunnyside Gardens, was a large expanse of potato fields. Abraham Levitt and his sons, William and Alfred, conceived of the opportunity here to build inexpensive homes on a mass scale to supply the new market. The entire plan was developed and financed by private capital, and the project entailed a self-contained community with its own infrastructure, that is, sidewalks, roads and utility lines. Previously, total community planning on this scale was considered impossible for private business, but Levitt changed this. In 1946-1947, 300 acres of Hempstead Plain were converted into the first "suburban" community as Levitt built small Cape Cod-style homes, which sold for $6,990. Levitt also introduced what we have come to call modular unit building. He trained teams of non-union workmen to specialize in one set of operations which they then performed on all houses. Despite subtle variations, the form of development required that all houses would be substantially alike. Thus, people were enabled to pursue an American ideal, a private home on Long Island.[4]

The demand was voracious, and was helped by FHA veterans' benefits. Levitt expanded the development, and by 1951 the community, originally named "Island Trees," but now called Levittown, included 17,544 homes. The Korean war-era houses sold for the famous price of $9,900. Other developers have stated in interviews that even at that time Levitt's net profit on each house was over 20 percent of the purchase price. Potential profits to developers per house remain roughly the same today, varying between 10 percent and 20 percent.[5] Twenty years later, the form of construction Levitt and Sons pioneered for developments has been widely expanded in the country. Levittown itself has continued to prosper. Houses are presently valued at well over five times their original purchase price, and many of the initial owners have long since departed for other areas after capitalizing on their investments by "trading up" to more expensive homes. Recently, a re-zoning proposal was introduced in the community to build a condominium complex of 329 units. The resistance of residents to this phase of suburban growth is great, and the proposal seems to have little chance of success. The development of Levittown, however, continues to ride the waves of metropolitan expansion as scarce housing will require higher density use to accommodate the growing regional population.

In Privatown there are several categories of businessmen who are engaged in developing land. There are builders, developers, and speculators. In addition, each type differs in their scale of operation as well as in their desire to switch hats and shift categories. The longer-range dealers in land are speculators who hold it as an investment in future growth. They generally buy up farmland and get it re-zoned to residential use. Dividing acreage into lots pays off extremely well once growth occurs and the individual parcels are sold. The initiative to a suburban development, however, commences with the developers' activities. It requires the assembling of a suitable tract of land as well as the construction of the necessary community infrastructure—roads, utility lines, sewers, street pavements and curbs. Developers are often speculators. They too must receive appropriate approval from the town board in the form of re-zonings, permits, or subdivision site inspections. Finally, a builder, who may also be the developer, completes the construction of homes. Township requirements regulate the builders' operations through the granting of building permits, certificates, fees or bonds. In all cases, each type of land operator requires considerable cooperation from the Privatown town board and its appointed agencies. In the following discussion we shall explore this relationship as well as examine the activities of the various developers operating in the township—the middle-range operators, spot builders, and national conglomerates. In a final section we shall discuss the effects of speculation and the activities of politicians and businessmen in the turnover of land.

MIDDLE-RANGE DEVELOPERS

Companies such as Ryan-Walker, Heather Acres, Elysian Homes, Dogwood Estates, and SPT Enterprises are locally-based, middle-range developers who build most of the homes in Privatown. They utilize the Levitt form to build on medium-sized tracts of land and supplying the community infrastructure. As developers, they implement their plan for land-use in the community and supply its roads and sidewalks as well as its houses. While this community planning is quite extensive, it is limited for the most part to the separate housing needs of the residents. Presently, it is neither profitable nor feasible under existing zoning for developers to provide open space or recreational centers in the communities they plan for and build. These amenities, therefore, must be provided by the township. Proposals for planned-unit development or new-town zoning are strongly resisted by residents.

Development-site requirements are modest in the area, because Suffolk County regulations do not require the building of sewers. Each house is equipped with a primary treatment cesspool. Drainage and runoff are handled by means of large fenced-in holes called "recharge basins" or "sumps." This is all the developers are required to provide. Recently, the absence of sewers has become an important environmental issue, and there is serious pollution of the ground water, especially in eastern Suffolk County.

Middle-range developers finance their construction through short-term commercial loans. Building on these medium-scale tracts never takes more than a few months. In order to cover expenses as rapidly as possible, sample or "model" homes are constructed. Prospective buyers are then received immediately, and in many cases, homes are purchased before they are completed. Areas undergoing rapid suburbanization ensure quick sale of such housing and enable middle-range builders to flourish because he can sell his homes within the time allowed to cover the short-term bank financing. This is called "leverage." In fact, once an operation has been effectively organized, a builder can function under high debt, use the banks and customers to provide most of the operating capital, and sell all his houses to realize a substantial profit, although he may not make much on the individual home.

Middle-range developers require considerable flexibility from Privatown's town board. They usually need some re-zonings on assembled parcels to build in quantity, apart from the permits and certificates necessary before construction can start. Delay on any of these prolongs the time of construction and restricts the developers' ability to operate under high debt. In Privatown middle-range developers are active supporters of the local political organization. They have contributed substantially to the Republican party's campaign fund. For example, informants stated that an important local developer, David Seymour of STP Enterprises, gave close to $100,000.00 for 1971 campaign financing. Informants also indicated that even developers who are Democrats and wish to operate in the township gave to the Republican party fund. Besides, several middle-range developers have very close connections with the party boss, Middleton, and his associates. On occasion, they have participated together in business deals, such as the development of a large townhouse complex located on a "limited-access" highway.

SPOT BUILDERS

If the suburban landscape is a result of the activities of private developers, the margins of that landscape, the spaces left vacant between the developments, are filled in by the spot builders. Unlike their middle-range counterparts, the spot builders deal with plots of land suitable for only one or two homes. Part of their business comes from people wishing to build a house on their own land. These are "custom-built" homes, although many of them conform closely to a model that the builder specializes in. However, most of the spot builders also attempt to acquire land owned by others, in order to build houses and then sell them to new arrivals.

Spot builders cannot exploit the economics of scale associated with tract development. However, they have a major advantage because they do not bear any site development costs. They work on land located in developed areas where the roads, sidewalks, and utility lines have already been put in by the municipalities or larger developers. But when the houses are completed, they can be sold at the going market price or the same price as the development homes. In an interview, a local builder stated that while middle-range developers make a maximum of approximately 20 percent gross profit on a house, a spot builder makes as much as 25 percent profit.

Many spot builders employ real estate agents who are thoroughly familiar with the local area. Using large street maps and town board records, the agents are able to compile lists of owners of vacant plots and sell the information to spot builders.

With this knowledge a spot builder can successfully construct and sell houses in quantities rivaling those of the middle-range developers. In 1971, Tic-Toc Homes, operating in a north-shore community of Privatown, built and sold 160 houses. Tic-Toc is an entirely local operation. Bliss Homes, a Smithtown-based firm, recently moved into the north shore of Privatown after exhausting the spot building opportunities in the former township. In 1971 it built approximately 200 homes.

While profits per house may range between $2,000 and $6,000 not all of the construction is of reputable quality. Flame Builders is well known for their particular brand of houses, which are often called "Sound Beach specials." They have achieved public disrepute, and have been known to begin falling apart after two years. Flame operates in the Sound Beach area, where residents are mostly old people, students at the state university, and young married couples. The company sells housing at a slightly lower price than other builders, and is therefore able to cash in on the lack of reasonably-priced homes or apartments. In the Sound Beach-Rocky Point area, one real estate agent handled 1,000 rentals and sales during 1971, with 40 percent going to university students and faculty, 20 percent to people on welfare, 20 percent to comparatively well-off people wanting temporary housing until their own homes were built, and 20 percent to middle-class families.[7]

Most of the time a plot of land located as undeveloped by a spot builder is less than the acreage required by current zoning ordinances for residential development. This occurs because spot builders operate in the older, previously developed areas of the township. Before construction can begin, the builder must apply for a variance from the Zoning Board of Appeals.

In the confusing terminology the appeals board does *not* rule on zonings but on appeals of previous zoning category requirements made by the town board, thus the term "variance." It is also significant that the appeals board is an adjunct of the town board, who directly appoint its members. Consequently, the board is a reserve of patronage positions available to the local political organization. In a recent case in which two appointments were made to the appeals board, no qualifications of the appointees were indicated at the public hearing. In response to a question about this from a local reporter, the township supervisor, Bibby, conceded that the selections were political decisions.[8] The relationship between the local political organization, the town board and the appeals board is illustrated in Figure 4 (Chapter 6).

Re-zonings, which are handled by the town board, represent changes in previously determined categories of land-use. Variances, on the other hand, do not deal with changes in category, but with minor variations in the building requirements within already existing zoning classifications. Building a house in a residentially-zoned area does *not* require re-zoning of the land, but if the house is a geodesic dome or less than the required size, it does not conform to the category requirements, and a variance must be obtained before work can begin. Variances are important because they generally involve a desire to build on plots of land that are smaller than what the existing zoning allows. In Privatown many sections were divided up into small units, such as one-fourth or one-fifth acre plots, before the 1950s, to serve the demand for summer cottages. Owners of this undeveloped land must get a variance if they want to build because present-day residential zoning requires minimum lots of one-third to one-half acre.

There is much pressure from local residents to prevent the development of substandard plots, because it radically increases the population density and may also lead to the construction of cheaper housing. The law, however, is now on the side of spot builders. In a landmark case, "Fulling vs. Palumbo," December 7, 1967, the New York Court of Appeals decided that if a piece of land has been held in "single and separate ownership" since being purchased prior to the uniform zoning laws, it may be developed by the single and separate owner at any later date. This ruling was later extended to any subsequent individual purchasing the land. To obtain a variance the "single and separate ownership" must be proved, which implies that the owner must show that he owns only this plot of land and intends to develop only it and does not own similar plots adjacent to it. This ruling is aimed to stop developers from buying up several nearby tracts and building large developments on the substandard plots. It is, however, a great incentive to spot builders who work on single housing lots.

As it appears, builders are in close association with the local political organization, because they rely heavily on favorable variance rulings. The spot builders, Heather Acres, Tic-Toc, Bliss Homes, and a few others, are known by informants to be heavy contributors to party funds.[9] This money is exchanged in a very *personal* way. The builders, for example, appear at party functions and buy blocks of tickets to the dollar-a-plate dinners. Records of these occasions are, however, not public.

Spot builders disrupt any attempt at comprehensive planning even though adequate land-use may be accomplished in certain areas. They fill in the margins of tract developments and make it appear as if one tract community merges into another. Furthermore, due to the current legal arrangements, variances are not usually reported to the town officials and do not always appear on the zoning maps. The township supervisor recently stated that

> certain changes approved by the Zoning Board of Appeals do not actually show up as such on the present zoning map.
>
> Therefore, a citizen viewing a zoning map would not get a clear picture of the zoning in an area he was particularly interested in.[10]

In fact, no one interested in planning for social services can get a clear picture from official township records of what areas are really like. Through the actions of spot builders, traffic congestion occurs without advance warning to new homeowners, and in some areas people seem to just appear from some invisible source. They load up the area and increase the burden on sewage drainage. A major problem for planners resulting from spot building occurs because of the unexpected introduction of new school-age children. Schools boards attempt to plan for their needs utilizing official township maps of new construction and census data. These figures are thrown off by the additional and unpredictable numbers of people brought into the area by spot building. One school district in Center Moriches has attempted to cope with this problem by sending out teams of people armed with polaroid cameras a month before the new school term. They are instructed to take pictures and mark down the location of new houses not indicated on township maps.[11]

The unsystematic growth of the area is caused by the ease with which spot permits are obtained. In recent years, applications for variances have greatly exceeded those for re-zonings. It is, therefore, easier for residents to keep track of the operations of several large developers than the hit-and-run activities of spot builders.

LARGE DEVELOPERS

Large developers include locally-owned companies such as Beckman & Sons and Elysian Homes, as well as national conglomerates such as

Kaufman and Broad, a Los Angeles-based firm, and Boise-Cascade, the nation's leading builder. As elsewhere, these corporations move into local suburban areas only when these have reached a more mature level of development. This insures a predictable return on investment. Their projects are of two types. Either they develop large tracts of land selling off houses in quantity, such as Beckman's Executive Fields and Elysian's planned unit development (PUD), Oak Forest, or they construct apartments and condominiums on small tracts of land developed in high density. The local companies in Privatown have concentrated their activities in the construction of single-family dwellings, whereas the national corporations specialize in apartments and condominiums. Both Beckman and Elysian have close connections with the township's political machinery. In the case of Oak Forest Development, Elysian not only worked with the local politicians but was also sponsored by the bi-county planning board (see Chapter 6). The national corporations, on the other hand, work through the local people to acquire the favorable re-zonings that are necessary for the high-density land-use associated with apartments. Thus, in December 1971, over the objections of the bi-county and local planning boards and local civic associations, Carmichael and Greenbush, a large national builder, received a favorable down-zoning of land to construct apartments. This was accomplished by having Walter Olsen (the law partner of Nelson, who is head of the Cotter Bank and a business associate of Middleton) represent them as attorney and Hans Kroeger act as the *original* petitioner. Kroeger is the owner of a local cement-supply firm and a regular associate of Middleton. Kroeger, upon receiving the re-zoning from the board, sold the land to Carmichael and Greenbush, realizing a substantial profit on the increased value of the land after the change of zone.

In the case of the local middle-range developers and spot builders, payoff to the dominant party occurs in an informal and personal way and is difficult to document. In contrast, the large developers have few direct dealings with the local politicians except through political donations. Thus, their transactions are easier to document. As an example, the Republican party boss owns a number of businesses which provide the opportunity for grateful individuals to transfer money through a formal commercial transaction. One of these enterprises is a local semimonthly newspaper called the Long Island *Mirror*. By buying advertising space, the large concerns can share their profits with the political leadership without having to involve themselves in the social activities of the local party as in the case of spot builders. A study done in 1972, in the newspaper's morgue covering back issues for a nine-month period, June 1971 to March 1972 (which was as far as back issues were kept by the paper), revealed that Beckman & Sons took out fifteen one-eighth of a page ads, thirty-nine one-fourth of a page ads and two full page ads. This amounted to substantial revenue to the paper. Furthermore, Beckman did not engage in advertising on anywhere near this scale in other more professional papers, or even the local community weeklies such as the *Three-Village Herald.*

In another example, as a condition for obtaining re-zoning, the developer of the Dollarhaven Mall, the Osage Corporation, was required by the Privatown town board to build two four-lane roads of substantial width to border their property, despite the fact that they run for only a short distance and do not appear to be necessary for traffic flow. The contractor for the roads, built at considerable cost, was a local company which, as *Newsday* discovered in 1968, was owned by one of the councilmen.

There are four basic ways in which money may be channeled to politicians and to the local political organization.

(a) By buying blocks of tickets to party functions;

(b) By purchasing a service from a business associated with township boss Middleton, such as advertising in the Long Island *Mirror,* or his automobile dealership;

(c) By engaging in a business deal with a corporation owned by one of the associates of the township boss, such as Seymour or Giannelli, or a business owned by one of the councilmen, such as Elam's construction firm;

(d) By campaign contributions from citizens and corporations.

Payments, furthermore, may be made to either the local political leaders or to the party, so that funds which go into private pockets also help to support the local political organization. This interchangeability of payments to party or leaders can also be illustrated by the fact that many campaigns are financed out of individual politicians' pockets. For example, it is widely believed that Middleton launched his first successful campaign with winnings received from the Irish sweepstakes.

The general pattern of association between local political leaders and developers is one of a limited relationship between politicians who seek funds to support the party and businessmen in need of a public decision to allow them to operate. In general, the politicians and businessmen do not have a continuing relationship, but cooperate only because public decisions are needed.

Real Estate Speculation and Banks: Parties in the Turnover of Land

Whether it is the Dollarhaven Mall's Argo Corporation, which is one of the largest commercial land owners in New York City and often in partnership with international financiers, or it is the spot builder, Flame, famous for its "Sound Beach specials," each developer plays a specific role in the growth pattern of the island. The decisions of builders, speculators, and politicians combine and set the tenor of development. Growth follows transportation, and this is the condition under which homeowners and other businessmen must arrive at their locational decisions. This nationally leading retail and housing region developed because of the locational advantages obtained through ownership of commercially strip-zoned highway land or the large tracts of flat, formerly farm acreage. The overall land-use pattern resulting from this process is one that would occur if there were practically no zoning at all.

The relationship between the select groups of developers, speculators, and politicians arises from a reliance on town board planning decisions for personal, profit-making, or party purposes. Most businessmen outside the construction industry are not part of this relationship and do not rely upon the Privatown town board rulings. Like the homeowners, however, they must contend with the social environment that arises from such actions. There are, however, other enterprises which play a significant role in forming the growth patterns of the region. These are the companies that provide the real estate and financial complement to the actions of the developers. It is our contention that the decisions made by these businesses are fundamental for the expansion of housing and the observed patterns of land-use. The remaining local businesses do not affect the pattern directly, but only fill in the spaces left by the combined actions of the more significant groups.

LAND SPECULATORS

Land speculation is so remunerative in Suffolk County that a group of investors has emerged as specialized speculators in real estate.[12] Most of them do not develop the land. Instead, they purchase it, assemble large tracts and plat it for future residential or commercial use. Just by subdividing the original purchase into sites suitable for single homes, they can realize their initial investment several times.

In Privatown, speculators push suburban development into areas that planners would prefer to remain in alternate use. They tend to ignore the region's needs for a diversity of housing and have been known to plat land which is poorly drained or at a great distance from power lines and other utilities. While the housing market must certainly expand to meet the growing demand in the submetropolitan region, it is unfortunate that speculators in Suffolk County have led the way in planning for that expansion. The economic interests of the speculators are so narrow that they feel no further responsibility once the land is sold, neither do they consider the long-term effects their actions have on regional growth. Morton Paulson has recently called attention to the many problems created by land speculators:

> Just as oil companies and franchisers scramble for commercial locations, leaving the countryside strewn with bankrupt gas stations and fast-food outlets, hordes of land hustlers race to grab up and subdivide huge tracts of acreage, regardless of housing needs. Often such profligate subdividing artificially inflates pieces of surrounding land, drives up taxes, causes disorderly growth patterns, and worsens housing shortages. Thus millions of people who have never had direct dealings with land companies are indirectly affected by their deprivations.[13]

In Privatown there are several types of land dealers who vary considerably in their individual financial situation and method of operation. In part this is due to the differing time scale they employ.

The first group are the "wildest" speculators, who attempt to predict development ten to fifteen years ahead. They generally buy up large tracts of vacant or agricultural land, subdivide it, and hold it as undeveloped until subsequent growth bears out their investment. As development proceeds, a second type of investor with a more limited time horizon steps in. He may have some characteristics in common with the first type, but his method of operation is distinguishable because he is more discriminating with regard to the tracts he holds, and because he prepares land for suburban development. The more limited time horizon of the speculators of the second type compels them to look for specifically valuable tracts in order to manipulate them through re-zonings or subdivision. Finally, as development of an area appears imminent, a third type of speculator will often enter the market. This type has the heaviest impact on the land because he makes his money be getting holdings re-zoned for specific purposes such as business or multiple-family housing. The higher intensity use which he obtains for his land greatly increases its value, and the speculator will then sell it to a developer or businessman who actually makes use of it. Thus, it is probably the last speculator who realizes the quickest profit on the growth of the area.

Some speculators may operate continually throughout the three stages. The initial holder of the land may not necessarily sell off to another, once development occurs. Suburban land-use, however, is already being decided upon as much as fifteen years ahead of actual development—and this is accomplished through the eventual actions of as many as three different speculators. A banker, active in financing commercial enterprises in Privatown, said in an interview that during the period of most rapid development, between 1950 and 1970, land in Suffolk County had already been parcelled out by speculators at least a decade before growth actually came to the particular area. At present, Beckman & Sons, for example, is holding a 2,000 acre site near Rocky Point which has already been re-zoned and subdivided into a residential housing area and which is waiting to be developed once the suburban tide reaches it. The site, if developed at present-day densities, could potentially house almost half of the 1970 population of Privatown.

Although developers play a different role than speculators, in Privatown they may be the same people. Sometimes, when the long-term profits on a tract of land appear to be considerable, the speculator will retain the re-zoned land and develop it himself. This has occurred, especially in a number of instances where shopping centers have been built next to major highways. Although there is a tendency to specialize in either real estate speculation or home development, different hats are worn by speculators whenever the promise of profits is great. An example of this is the group of Atwill, DeSilva, and Von Cleef, who have operated in Privatown under thirty-eight different corporate names and won a total of eighteen out of twenty-three re-zonings between 1962 and 1968.[14] Over the years the group has been involved in several business deals with at least four of the seven township councilmen. The 1968 zoning scandals involved the two town board members Puglia and Elam and the above group. The councilmen were subsequently indicted on conflict of interest charges. Their activities were made somewhat difficult for the public to trace because of the many names that the group uses in applying for re-zonings: Yoyodyne Realty, Piedmont Realty, Cornerstone Realty, Sybil Associates, Cow Path Realty, Highway Ventures, Atcleef, Atmac Realty, Calverton Associates, Ocean View Associates, etc. Citizens appearing at town board public hearings, while possibly knowing of the group, cannot keep track of all these other labels, some of which sound quite innocuous. In fact, the use of names, such as "Elysian Fields" or "Heatherwood Estates," is commonly adopted to convey a pastoral image that obscures the developed nature of the area.

An illustration of the way in which speculators also operate as developers, and in which their activities could successfully go against the wishes of local residents, can also be given using this group. Atwill, DeSilva, and Von Cleef assembled a large tract of land in the Center Moriches area. As development proceeded rapidly, they decided to retain the land and use it for their own business purposes. They got it re-zoned for a commercial shopping center in spite of the objections of *both* the local residents and the township planning board. They kept the newly re-zoned land and built Moriches Shopping City.

One Atwill owned corporation, Island Horizons, operates consistently in all areas of speculation and business development. It holds land and, in addition, owns gas stations, shopping centers and multiple-family housing throughout Suffolk County. In the 1968 scandals it was discovered that Middleton and councilman Elam assembled a parcel, got it re-zoned by the town board on their *own* motion, and then sold it to Island Horizons as a site for a gas station. Middleton and Elam made $64,000 profit on that particular re-zoning. This is one of the clearest cases in which public officials and businessmen play interchangeable roles. Elam and Middleton, the public officials, operated as speculators and sold property to other speculators operating as gas station owners. The pattern of interchangeability and close cooperation was evident in this deal, because another town councilman, Puglia, was a silent partner with the Atwill group in Island Horizons.[15] Local government and the local businessmen are sometimes indistinguishable.

In addition to operating closely with the zoning decision-makers in the town, speculators require a close working relationship with town tax assessors. Once again the individuals involved often play all roles. Tax assessors are responsible for making crucial decisions on the taxes to be paid on speculators' holdings. These taxes are a critical variable in the speculators' time horizon. There is ample evidence of close cooperation between these speculators and tax assessors in Privatown. Profit sharing mechanisms between town assessors and speculators take on a variety of forms from cash payment to the transfer of portions of the land holdings. Thus the assessors can also become speculators. In 1954, for example, it was revealed that the town assessor had himself operated with a group of speculators and sold a tract of land, making $92,000 in profit.[16] In 1959, this assessor, Arthur Callia, was indicted in the land-grab scandals and pleaded guilty to charges of bribery and conspiracy.

The continual disclosures of land grab scandals during the 1950s and 1960s indicate that not every tax assessor, local official, or party functionary works in dishonest conjunction with land speculators.[17] However, our point is that a small number of these individuals do, and their actions seem sufficient to change the pattern of development.

Due to the rapid growth of Privatown, the proliferation of this group does not heavily burden the residents nor arouse them to much resistance. Most residents seem well satisfied with home ownership as a source of equity and have gained substantially from appreciation. Local government has played an essential role in this process by channeling development to preserve and enhance property values. The local political leadership and their business associates have also profited from their control over zoning which determine the extent and location of high-yield commercial and multiple-family land-uses in the midst of affluent suburban communities. The majority of Privatown's people have all speculated, in a sense, on its future and directly benefited from township growth. Those that did not share in the returns on investments, the newer residents, minority groups, and the less affluent elderly, will find little future relief as the social costs of such speculation are realized.

The financial and speculative linkages in Privatown can be illustrated better by examining other parties in the turnover of land—the financial investors and banks.

FINANCIAL INVESTORS AND BANKS

Speculators desiring to cash in on the region's growth require large sums of money in order to operate over several years. For this reason they work with financial investors. The major source of financial capital to support speculation is found in local banks, both commercial and savings and loan, and banking is a dominant sector of the local economy.[18]

The Nassau-Suffolk region is considered *first* in the nation in expendable household income. This large pool of money is a source for local banks, many of which are owned by conglomerates. While most banks may not necessarily be directly associated with the developmental pattern, they do play a significant role. The entire spectrum of private decision-makers from the individual local home owners to the commercial developers rely on banks to supply them with anything from short-term ninety-day loans to thirty-year mortgages. The gap between economic activities and their costs is filled in by the liquidity supplied by the local bank.

Several banks are directly associated with speculators and local politicians. One example is given by the activities of Donald Dumbrille, the chairman of the board of Benjamin Franklin National Bank. He is involved in a large speculative combine which has assembled a tract of land in the undeveloped eastern region of Privatown. Six partnerships with 41 persons involved in some or all of them have paid a total of $2,115,870 for over 350 acres.[19] The attractiveness of the site is due, in part, to the fact that it is adjacent to an airfield used by the Grumman Corporation, and the bi-county master plan indicates that the area may possibly be turned into a fourth New York jetport. At the time of the assemblage of the tract, carried out under the different partnerships so as not to arouse the suspicion of local residents, Drumbrille was also a member of the bi-county planning commission, the agency responsible for drawing up the proposed master plan. After the *Newsday* disclosures of this conflict of interest, however, he was forced to resign from that post. Thus, while no direct information on the use of bank funds is available, the connections between bank officials and real estate speculators may have significant consequences. When we add to this the county post, the role of Dumbrille seems to indicate the possible connection between the sources of capital and the proposed use of money for speculation on the future growth of the area.

Sometimes banks have been known to act in direct cooperation with other commercial and political interests in financial deals. Newer banks need the cooperation of other businesses and the political leadership in order to build their cash assets. The Cotter Bank, formerly owned by local businessmen but now controlled by a major national conglomerate, is a good illustration of the process by which local banks can obtain upward mobility. In the 1960s, while this bank was still locally owned, its chairman, Morley, belonged to the Privatown Town Industrial Commission, which was entrusted with the job of enticing new industry into the area. Several of the commission members were associates of Middleton, and also active land speculators, as this was a convenient and legitimate way of using the public trust to enhance their positions.[20] When the industrial commissioner decided that the township needed a supply of low-cost land which it would then lease or sell to attract small industry, Morley volunteered to set up the land bank as his personal nonprofit service to Privatown. He was thus placed in a position of decision-making that held extreme importance for the future growth of the area.

Morley began assembling a 100-acre parcel for an industrial park, and the various township agencies threw their money and full support behind the project. Thus the town board: (1) down-zoned twelve acres of the parcel *on the board's own motion;* (2) expended township funds promoting resale of the land on a top priority basis, and (3) promoted the first resale by approving a low cost State Job Development Authority loan for a firm which bought a part of the parcel.[21]

Two small industries were brought into the tract, and the feasibility of the project became manifest. At precisely this time, Morley *sold off* the remainder of the land to a private developer, thus selling out the public interest and pocketing the profit on the land sale in his own bank. *Newsday* estimated that Morley and Cotter Bank made $263,156 profit on the deal.[22]

While banks are not in the forefront of suburban development, they are, nevertheless, indispensible to speculators and developers as sources of funding. In some cases, however, they actually use their position as suppliers of capital to the local political leadership and real estate speculators to operate on their own initiative. Once again, a distinction must be made with regard to these activities. Not every bank, bank official, real estate holder, or local politician is involved. We seem to find, instead, that a few individuals within these groups recurrently participate in the decisions which guide the growth processes of the township. Thus, in the Morley case, it was his initiative and close work with the political leader of Privatown that created the industrial park. Those businesses which then moved into the area received the locational advantage of being in the park but none of the initial benefits of the turnover investment in land through zoning changes.

In addition to banks, many private individuals become involved in speculation as financial investors. In Privatown, this is the case with a number of national and international personalities who are partners with speculators and politicians in corporations active in developing choice real estate packages. Carl Middleton often works in partnership with famous sports figure Lawrence "Skip" Trendleberg on his land deals. The latter supplies the capital to take advantage of opportunities created by political control. It has been stated by some informants that international financiers and even former U.S. presidents are often found as associates in similar relationships operating in Privatown and throughout Long Island. A most interesting case concerns William T. Ronan, the head of the New York-New Jersey Port Authority and formerly of New York City's Metropolitan Transit Authority. During the confirmation hearings conducted by the Senate on Nelson Rockefeller's appointment as vice president in 1974, it was revealed that Ronan was the beneficiary of over $600,000 in loans from Rockefeller which were later converted to gifts. During that same period, as reported by a local newspaper, Ronan invested heavily in Long Island real estate and improved his net financial position by over $400,000.[23]

In concluding our discussion, the following points can be made:

1. The local political party and those businessmen involved in submetropolitan growth tend to merge into something of a land development corporation. The interests involved in this select group varies from the use of town board decisions by politicians for personal gain and party support, to profit-making by speculators, developers, and investors. The cooperative relationship existing between businessmen needing public decisions and the political leadership involves a variety of techniques for the exchange of money. Businessmen either make political contributions to the party or to individuals, hire a service performed by a business or legal interest of the political leadership or associated with it, or are involved in a business deal with politicians for mutual benefit.

The profit-making interests in the township are divided between large national concerns and local businessmen. Cotter Bank is an example of a purely local operation that eventually was bought out by a national bank. The major defense corporations and building conglomerates, which operate throughout the country, are examples of national concerns. Speculating combines have been known to include township boss Middleton and local financial investors or speculators as well as international financiers. The roles of local men vary from one development project to the next as they shift from politicians to businessmen to speculators and back again. Regular businesses needing favorable decisions from the local political organization support the party in some way. Local corporations tend to exchange money in a personal way, while the dealings of national companies tend to be more formal in nature. Thus, when a major oil company wishes to obtain a favorable re-zoning for a gasoline station franchise, it uses a more direct means of monetary exchange. Connections between all layers of government and the private sector testify to the incredible variety of ways that money can change hands between parties interested in the turnover of land.

2. The decisions made by the politicians, speculators and housing developers lead to the same land-use pattern as would result from no planning or zoning. Businesses not directly dependent upon regional development, such as industrial plants or department stores, and homeowners must locate in the environment created by the actions of this select group. While the latter receives the major benefits of rapid growth, the social costs must be shared by all.

3. In the initial stages of rapid suburban development, however, residents have not been hurt badly by this pattern of growth. Most individuals involved have given primary consideration to short-run profit-making and

privately oriented interests, so that even homeowners have speculated on their investments with respect to future growth. In the past, the actions of politicians, lawyers, investors, speculators, businessmen, as well as homeowners have complemented each other to produce the familiar pattern and interests associated with suburban development. Leonard Downie, Jr., has summarized this view on a national basis:

> The suburbs have sprawled inefficiently, because the farther out the speculators go into the countryside the cheaper the land is. The natural environment has been defiled, because more careful development would cut into profits too much. Houses are free standing and lots wastefully big, at the expense of parks and other open spaces, because that brings homeowners more profit on resale. Individual homeownership is also encouraged by preferential tax treatment and government backing for home mortgage loans. Most of the houses are built cheaply by developers cutting corners to maximize profits, yet they still sell easily to buyers who expect to change homes frequently—"buying up" to a better house with the profits from the sale of the last one—and who are more interested in short-run profit potential than a home's lasting value as part of the community's stock.[24]

Today, however, homeownership speculation and the substantial profit-taking of politicians and businessmen seem less assured in Suffolk County. Metropolitan expansion has declined and appreciation of house value, while still rising, has slowed in rate of increase. Until the cost of living in the Nassau-Suffolk region matches that of the central cities, suburban patterns of development and the movement of people into the area will persist, but at this slower rate.

NOTES

1. Cf., Chapter 1 for a review of federal housing and mortgage subsidy programs which form the basis for government involvement in social planning.

2. Leonard Downie, Jr., *Mortgage on America,* New York: Praeger, 1974.

3. Morton Paulson, *The Great Land Hustle,* Chicago: Henry Regnery, 1972, p. 66. Paulson claims this long-range devouring of undeveloped land exists also today and on a massive scale: "According to the best estimates, Americans in the early 1970's were contracting each year to buy property priced at more than $5 billion—vacation or second-home lots, retirement homesites, building lots in promotional subdivisions . . . in which lots are sold long before there are people around to occupy them." Ibid., p. 2.

4. An analysis of Levittown, summarizing its growth progress to the present, was made by *Newsday,* "Levittown: 25 Years Later," Part II, 7/9/71.

5. The Kaiser Commission report on housing estimated that suburban developer profits averaged 12 percent, cf., *A Decent House: Report of the President's Committee on Urban Housing,* Washington, D.C.: Government Printing Office, 1968.

6. New York *Times,* BQLI section, 12/7/75, p. 3.

7. Evidence on Flame Builders was obtained from a local real estate agent who rents and sells most of the homes built by the company.

8. *Newsday,* 12/8/71.

9. Information on the cooperation between spot builders and the dominant party leaders was supplied by a local attorney who handles variance cases for some of the builders. In an interview he explicitly stated that campaign contributions, however difficult to disclose fully, are used as a profit sharing device in Privatown. In fact, he stated that if it were possible to discover the exact amount of the contributions made by each spot builder, the ranking among them would correlate completely with the magnitude of the volume of construction each builder carried out over the previous year's time. Due to the variety of techniques utilized in giving money to political parties, this correlation cannot, however, be explicitly carried out.

10. *Three-Village Herald,* 10/11/72, p. 2.

11. School teachers and officials interviewed in the rapidly growing sections of the town declared that measuring the impact of spot builders was a definite topic of discussion in school planning sessions. Polaroid picture taking is one technique used to keep up with the changing physical reality of rapid suburban development.

12. Recently, speculators have been scrutinized by several critical studies because of "epidemic" proportions in metropolitan development. In addition to Downie and Paulson, op. cit., cf., Bruce Lindeman, "Anatomy of Land Speculation," *AIP Journal,* (April 1976). The reader is, of course, referred to the most original and, perhaps, insightful study of speculation: Henry George, *Progress and Poverty,* written in the 1870s and reissued–Garden City, N.Y.: Doubleday, 1926.

13. Paulson, op. cit., p. 5.

14. This information was obtained through research at the *Newsday* morgue. Thus, while all data are from news clippings, the date and pages of those clippings were not always indicated.

15. Ibid., May 8, 1968.

16. Ibid., May 4, 1958, p. 4.

17. Cf., *Newsday* coverage of land scandals for details on the individuals involved.

18. Of the studies done on speculation, few have tied in banks. Downie, op. cit., does, but more work is needed in the area, especially in connection with central city as well as suburban submetropolitan speculation.

19. *Newsday* morgue clippings.

20. Ibid., 12/16/68.

21. Ibid.

22. Ibid.

23. New York *Post,* 10/18/74, p. 1.

24. Downie, op. cit., p. 87. The author calls suburban regions "profitopolis," because of the extent of this speculation.

Chapter 6

PROFESSIONAL PLANNERS: The Limited Success of Rational Land-Use Decision-Making

At first glance, it may seem odd that a study of social planning places a discussion of professional planners so close to its conclusion. By this time, however, we have been made aware of the limited role the profession plays on Long Island. Land-use decisions are made by local governments and by businessmen. The interests of local residents, housing developers, real estate speculators, and politicians in Suffolk have combined to produce a pattern of strip zoning and uninterrupted development called spread city. As Robert C. Wood has observed, local suburban townships have "rolled with the punch of urbanization," rather than controlled it. In a sense this low-density sprawl pattern is the very opposite of social planning. It represents the aggregation of many private interests and short-sighted politically expedient decisions, instead of the articulation of broad, future-oriented community goals. Apparently, government planning programs are merely a facade for this privately directed process. For example, the former executive head of Suffolk County, Frank Laughton, has recently commented on the perplexing way socially useful planning principles seem to have been systematically ignored in township land-use decisions:

> Discount houses, hot-dog stands and gas stations seem to spring up around the county without reference to any comprehensive plan. The facts of political life are clearly indicated by concrete evidence that possibly the recommendations of the professional planners are being ignored or shelved, that the backroom zoning and political blue printing goes on as before behind a headline front of professional planning.[1]

In previous chapters we have explained this apparent mystery of limited planning through the negative effects of political control. It is experienced by the political leader of the county and by local residents as a paradox, because of the existence at all societal levels of numerous planning agencies involved in the process of social planning. But if planners do not implement land-use decisions nor guide directly social growth in our society, we are left with the intriguing question—What, then, do planners do? In Suffolk County professional planners are involved primarily in the task of producing and upgrading a master plan. The master plan is a comprehensive guide for social growth which is submitted to the county and the local townships on an advisory basis. It involves the evaluation of demographic, business, housing, and other socioeconomic trends in the region, and, then, details a comprehensive description of the preferred land-use pattern for future growth on the basis of that information.[2] It aims to outline the "rational" steps which can be taken by local government to control development for municipal purposes. According to this view, planning elevates

> the quality of democratic politics by broadening the perceived range of possible solutions, by pointing out ways in which specialist proposals affect each other, and by suggesting means of harnessing the techniques of modern specialists to serve broad visions of the public interest in concerted fashion.[3]

The theories of planning used by the profession in the United States substantiating the master planning function concern themselves with aspects by which "rational" decision-making can be formulated for political leaders to control social growth. Such an approach assumes planner advisory status and concentrates on principles of design which emphasize the control over land-use and construction, that is, it stresses design of the physical form within which social activities take place rather than attempting to analyze or control the activities themselves. This is reflected in planner reliance on zoning ordinance, municipal building, and environmental codes as the major weapons of the master plan. The comprehensive suggestions made to the political leadership aim at preventing any piece-meal, limited attempts by them to use municipal control of land-use and public works, such as highway construction, on an ad hoc basis in response to social growth needs. It almost goes without saying that such planner recommendations do not assume the possibility that these powers of political control will be used for political purposes or personal gain. The master plan, therefore, provides a "rational" design to contain future development and assumes that social activities themselves will not be planned, but, instead, be coordinated by the political process and by the business goals operating in the private marketplace.[4]

The primary characteristic of master planning in Suffolk County is that planners have little power beyond the persuasive credibility of their recommendations. They advise local government regarding suburban growth trends. However, the latter retains the power of implementation–government makes the decisions. In our society there is, therefore, a separation of planner and political functions. At the most local level of the suburban township the planners are political patronage appointees. This feature of political control over the planning process is common to all levels of government and seems to be favored by the polity as part of representative democracy. In Suffolk County, however, decentralized home rule has led to several social problems. Local control of land-use has produced a sprawl pattern of growth, has not been able to preserve environmental quality, and has led to racial and income segregation on a regional scale. Professional planners in Suffolk have recently been concerned with these problems and they are presently seeking greater public authority and control over planning implementation. Planners, for example, have begun agitation to overcome the inequities of racial and income segregation. They are also concerned with ecological considerations and the need for more moderate and low-income housing in the region. In pursuit of these issues planners are becoming active participants in the political decision-making process. In short, Suffolk County planners have become politicized as an interest group which attempts negotiation, at times behind the scenes, with businessmen, local political leaders and township residents for support of planned land-use recommendations. The "rational" view of planners has, therefore, become one of many arguments debated in the political bargaining process for control of township resources, and planner-supported public issues have increased in frequency over the past several years.[5]

In the following discussion we shall present several case studies of professional planning in Suffolk County and Privatown which illustrate planner response to their lack of implementation powers and the limited achievements of planner "rationality" in the political bargaining process. The lack of political implementation power, however, is not the only weakness of the professional planning approach. Master planning is limited by at least two other shortcomings which greatly affect the ability of government planners to realize their goal of "rational" guidance for social decisions. These include, first, the commission of the physical fallacy and a reliance on a technical, limited notion of "rationality," and, second, the presence of what we call the elitist-populist dilemma which greatly restricts the ability of planners to gather wide public support. Before reviewing our case studies, therefore, let us briefly discuss these weaknesses of the professional planning perspective.

The Weaknesses of Professional Planning

THE PHYSICAL FALLACY

Physical planning concerns itself only with the proper allocation and use of space. This is done under the implicit assumption that adequate living and working arrangements for society can be achieved through the use of construction and landscaping technology, while taking the relations and mediations of those processes for granted. Such a perspective follows from the tradition of planning in the United States since the turn-of-the-century City Beautiful movement. The problems and processes of urban growth are viewed primarily in terms of design. Planners today are taught, for the most part, to follow in this architectural tradition. The director of the combined Nassau-Suffolk Bi-County Planning Board, for example, was trained as a landscape architect.

As Herbert Gans has shown, this approach to planning assumes that social and cultural patterns of interaction can be successfully manipulated merely by proper designs of physical environments.[6] This is an example of reductionist logic, and it is a fallacy. The physical approach to planning is a special case of what can be called technical rationality, the adoption of efficient means for the attainment of a given goal. Technical rationality always leaves the goal of decision-making open-ended. In drawing up the master plan for the development of Long Island, for example, the bi-county planning board analyzed the socioeconomic patterns of growth and projected the future trends of those patterns. Utilizing this information the plan then presented the most efficient form for these development trends. It did not, however, attempt an evaluation of the activities themselves, nor did it specifically plan for ways in which the social needs of the developing region created by the trends could be met. Confined to this limited theoretical perspective, the recommendations of planners can be considered to be "neutral." Technical rationality has this property. The social, economic and political processes of development are merely given a more efficient and suitable form so that the goals of growth remain open-ended to be chosen by the operation of these processes. The developmental process, however, contains definite judgments on resource allocations and values which are *not* neutral. As Altshuler has observed:

> [The planner] . . . controls so little of his environment that unquestioning acceptance of its main features is a condition of his own success.[7]

At its best the planner's perspective is a rather naive view of social development. It ignores, for example, the actions of speculators, the effects of racism and exclusionary zoning, and the incentives created by government programs, such as FHA home loan mortgages, on trends in the region. It takes for granted these processes, instead of subjecting them to "rational" analysis as well. Paul Diesing has pointed out that there is, in fact, a taxonomy of rationalities that must be dealt with in viable social planning.[8] Social, political, and economic considerations are as important to societal decision-making as are the technical, engineering aspects of landscaping patterns of growth.

THE ELITIST-POPULIST DILEMMA

The professional planning approach also embodies a mixture of elitist assumptions that the planner is the expert in knowing best how to control growth, along with populist beliefs that there is an easily aggregated "public interest" which planning can fulfill. This perspective does not entertain the broader question of the public participation in the selection of goals, the search for alternative choices, nor the adequate monitoring of outcomes in the transition from paper scheme to practice. This is an elitist approach. On the one hand, planners feel that they should be entrusted with the responsibility for such decisions as a profession in the same way that patients are not called upon by surgeons to participate in determining the course of an operation. On the other, however, the planners, unlike the surgeon, are limited in their ability to implement their scheme. They need the public's support for their ideas. They *must* include local resident participation in the planning process at some stage in order for the master plan to be accepted by local government. This presents planners with a dilemma. The courting and attempted acquisition of citizen support is accomplished by a public relations effort and through the use of pseudo-forums. The latter are public "information" meetings that generally involve panel discussions and speaking engagements by the professionals on the aspects of the plan. We call this activity "explanation" and it is discussed more fully below when we consider the content of the bi-county master plan.

This position is somewhat analogous to the "good government/progressivism" attitude exhibited by local populist reformers. The professional planners wish to lead the polity by virtue of their expertise as "seers" of the suburban future. In practice, however, elitist-populist planning often ignores the needs of local residents. The public is allowed to participate in the planning process only through the exercise of a political veto power as the planners, local political leaders and businessmen monopolize the initiative. The recent controversy over the Oak Forest planned-unit development in Privatown is an example of this public position, and we shall examine it in detail as a case study below. Here the local residents were faced with accepting the plan in a series of information meetings or fighting the proposal in the re-zoning process. No attempt was made to include the citizens in a discussion of alternatives to the plan nor to examine their divergent interests.[9]

In trying to understand the weaknesses of the professional planning perspective and its inability to develop an enlightened notion of viable social planning, we have come to recognize that it is constrained by a widespread, persisting belief in a self-governing society with automatic mechanisms of social adjustment. These beliefs are embodied in the ideology of free enterprise. Planners for the most part accept this view and the social environment in which they work. Their appeal to be included in the privately oriented social decision-making process centers around the proposed benefits from the government use of their elitist expertise. Ironically, however, the weaknesses of their approach work against the public acceptance of a larger role for planners and contributes in no small measure to persisting citizen belief in the superiority of automatic mechanisms of adjustment over planned growth. Because professional planners often ignore the interests of local residents, there are *real* fears present here. The public may be correct in its belief that giving planners direct powers of implementation could result in totalitarian controls, and they are, therefore, more willing to take their chances with the present system of political control which separates planners from the power of implementation. The urban renewal literature, for example, documents the results when planners and local political authorities were given broad powers over community growth, as does the extensive record of planner Robert Moses in New York City.[10] The persisting public fears of a planned society are as much a fault of the weaknesses of the professional planning approach as they are of the way in which government itself has taken advantage of the concept of planning in order to set up and enlarge its bureaucratic apparatus of social control at the expense of local citizens. We shall return to such issues raised by planning limitations and the problems of political control in our final chapter.

In Suffolk County there are two agencies, one public and the other semiprivate, which attempt to provide comprehensive guidelines to its townships. These are the Nassau-Suffolk Bi-County Planning Board and the Suffolk Community Development Corporation. We shall discuss the operations of both of these professional planning agencies in a series of case studies. We hope to illustrate the limitations of planning in the suburban region, including the lack of political implementation powers, the physical fallacy, and the elitist-populist dilemma.

The Nassau-Suffolk Bi-County Planning Board

From 1945 to 1958 the economy of Long Island expanded due to the post-war prosperity and the involvement in the Korean conflict. Government institutions and defense-related industries were the major areas of employment in the region. In 1953, for example, the five giant defense corporations, Sperry, Arma, Republic, Fairchild and Grumman employed 67,394 people. This was the backbone of the island's economy.[11]

The recession which occurred during the Eisenhower administration, and which affected the entire country, hit Long Island especially hard. Suffolk was hit harder than Nassau and was inflicted with the inordinately high rate of unemployment of over 12 percent during those years. Nassau residents, due to their close proximity, could depend upon New York City for available jobs to a greater extent.

The origins of the bi-county planning board are immersed in these trends. The worsening economic condition created a need on the part of the business community to utilize expert technical assistance in stimulating the economic growth of the region. The concept of coordinated regional planning, however, arose through the combined efforts of the first Suffolk County Executive, Frank Laughton, and his Democratic ally, Nassau County Executive, Eugene Freeman. Laughton was first elected in

1959 in the wake of the 1958 Republican land-grab scandals. He was a civic engineer with twenty-five years of county experience and a strong interest in land-use controls. He labeled his perspective as one of non-partisan professional planning and won the election for the Democrats. As county executive he formulated massive social programs to get Suffolk "moving again." At the time Laughton and Freeman controlled their respective counties, there existed two separate planning boards appointed by each county government. Steven Walker headed the staff of the Nassau Planning Commission while Larry Pendleton was the counterpart in Suffolk. The two Democratic county executives met with little success in getting their respective programs approved. Laughton, in fact, was faced with a Republican voter revolt against his spending programs in the first few years of office. The county executives, however, picked up some allies among the businessmen who were worried about the island's slumping economy. This was especially true of the banks. The businessmen began to think in terms of a combined planning board and regional planning agency. The news media on Long Island, such as *Newsday* and the Long Island *Commercial Review* also came out strongly in favor of two-county planning.

During the worst part of the recession, in 1963, a resolution was passed by both county legislatures setting up a joint agency of the formerly separate Nassau and Suffolk county planning boards. This bi-county planning board had an enlarged budget, and its original purpose was to investigate the present growth trends of the island and suggest ways of combating the recession and managing growth. Larry Pendleton, a landscape architect, was given the job as executive director at $10,000 a year and a staff of thirteen people including the members of both county planning boards. The bi-county board itself consisted of its chairman, Gerald MacReady, a well-known businessman and a member of the board of numerous corporations including Gulf and the Beneficial Life Insurance Corporation; Stanley Leonard, the vice chairman, who was an attorney; Robert Lawrence, the director of Security National Bank; Jerome MacLane, the owner of an automobile dealership; Donald Dumbrille, a director of the Franklin National Bank; and Walter Simpson, an attorney. Dumbrille subsequently resigned in February 1970 when *Newsday* disclosed he was also a member of a large speculative combine operating in the area of the bi-county's proposed jetport at Calverton, Long Island.[12]

The bi-county planning board, thus, represented a curious mixture of Democrats intent upon the expansion of the region being guided by professional planning and engineering principles, and a solid component of

Republican-affiliated corporate interests. When his political power declined in the elections of 1963 and 1965, Laughton discovered that the stumbling block to his long-range plans was the orientation of ordinary residents, who supported the ideology of local "home rule." This belief gave the power of zoning to the politically elected town boards. Banking and corporate components of the business community, therefore, found themselves aligned with the Democrats in an attempt to stimulate the economy in those recession years, despite the Republican orientation of restricting government spending and a belief in local government control.

Shortly after the board's inception, the economic picture changed as consumer demand was given a boost by the Kennedy-Johnson tax cut of 1964. The defense establishment obtained a reprieve from recession woes with the immense escalation of United States involvement in Vietnam. By 1965, the prospects for Suffolk County's prosperity were excellent. In that year, Governor Rockefeller announced that funds were appropriated to extend the Long Island Expressway all the way through Suffolk County to the town of Riverhead at the east end of the Island. This construction opened up the region to rapid suburbanization as the defense-oriented economy boomed.

As a consequence of the shift in historical events the bi-county planning board became strongly identified as the child of the Democratic county leadership. From its inception, the planners were given an advisory role. In the traditional view of planning in this society, implementation powers of land-use control remained in the hands of the local township governments.

In 1965, the bi-county board received a one and a half million dollar grant from the Federal Housing and Home Finance Agency. The express purpose of the support was to aid in the drawing up of the bi-county master plan. During the next five years, between 1965 and 1970, the board confined itself to this task. In addition to the plan, spin-off studies on public transportation, employment and income trends, community redevelopment schemes, and a massive land-use inventory were published.

The Democratic leadership of the county encountered stiff resistance to its programs as affluent Long Islanders called for minimal taxation and the limitation of migration into the region. Throughout this period, Republican localities reiterated their belief in private enterprise control of housing and local home rule. County Executive Laughton was a virtual figurehead as contemporary trends made his programs to improve the economy of the region patently irrelevant.

In 1970 the master plan was unveiled with much public fanfare. *Newsday* published a special issue on it, and Pendleton immediately embarked on a speaking tour to explain it. The plan itself was printed in summary form on high gloss paper with a multi-colored scheme and was widely distributed to the public. Its content was a classical example of physicalist logic applied to the control of land-use. While some social goals were also indicated and expressed as functions of the projected increase in population, the public participation in the determination and implementation of those social goals was ignored.

The plan was only a paper scheme. In order to increase the political clout of the bi-county board and simultaneously aid the development programs of the county executive, Laughton and Pendleton adopted a strategy of working closely with each other. The period which followed was marked by the planners' search for implementation powers aided by this political alliance.

Laughton announced that the board would be funded as a permanent bi-county agency and would pursue the goal of implementing the land-use recommendations. With this political backing the head of the planning board, Pendleton, could now enter into the local township arenas and negotiate with the local political organizations in order to influence the course of development. An analysis from 1970 by an observer of county trends indicated this new direction:

> Up to now, the board has concentrated on the preparation of land-use maps, analyzed the economic conditions and put a magnifying glass on just what is happening in transportation, public facilities, and housing. In the future, the board's role is likely to focus more on implementing the comprehensive plan that grew out of all the research, although implementation formally lies with the counties, towns and villages.[13]

Conceived as a joint project by the Nassau and Suffolk county boards of supervisors, and financed by a grant from the Johnson Administration, the board in 1970, became a vehicle for an attempted erosion of the Republican local power base and simultaneously a way of breaking up the confines of the traditional role of planners. Rather than an overt conspiracy of Democrats, the board's activities in Suffolk County need to be viewed in the light of a symbiotic union of political maneuvers to erode the Republican party's home rule political base and the entrepreneurial efforts of non-partisan Pendleton to redefine the traditional role of planning and planners in the community.

THE CONTENT OF THE PLAN

The plan incorporated the traditional assumptions and propositions of technical land-use rationality utilized by professional planners. The primary assumption of this approach is always resource scarcity. As stated by the master plan itself: "The plan is responsive to the future demands of the population and reflects the fact that the natural environment is not limitless. The number of people who can be accommodated is limited by environment constraints, transportation, and the need to preserve open space and shore fronts for conservation and recreation."[14]

From this base three propositions were developed which were envisioned as holding the key to the proper development of the island: Development should utilize the principles of *corridors, centers,* and *clusters.* To begin with, the island was viewed as exhibiting a series of corridors running east-west. The center was a transportation-industrial spine, containing the main branch of the Long Island Railroad and the Long Island Expressway. The plan recommended that all heavy industry be confined to this corridor. Bordering on either side were the residential corridors. The wealthier residential and recreational areas were to be located at the shore fringes.

Along with *corridors,* development was to take place in *centers.* Rather than having sprawl by functional units, shopping, administration, recreation, and services all were to be located in "activity centers," advantageously placed to service communities and separated from other centers by open space.

The third concept was *clustering.* It called for the construction of housing in higher densities (one-third and one-fourth acre plots) than at the previous level (one-half acre). The left-over land would then be kept in the development as open space. In this way, it was hoped, the suburban sprawl of tract homes would be avoided.

In addition to these concepts, goals reflecting the needs of the projected population were conceptualized. These were associated with housing, transportation, and recreational use. Interesting to note, however, was the fact that the plan for land-use was not in any way integrated with these future goals of growth. The bi-county board acknowledged some of these population needs, such as calling for more low-income housing, but *no plan* for the adequate realization of such social goals was discussed. In a sense, the treatment of future resident needs by the master plan was even more misguided than the physicalist proposals. The latter were developed and embodied in the glossy paper scheme distributed to the public, who consumed it only as an attractive form. At the same time, the public knew that the board had limited political powers to implement the plan. The social goals, however, were based upon an extensive demographic analysis of *future* population needs. The board had said that the Nassau-Suffolk region would *require* the additional transportation and housing facilities in the future, if previous growth trends continued. The immediate and almost unified reaction by the public was to do all they could *not* to let those trends continue. Thus, in the case of future social needs, the public responded with a "plan" of its own to limit population growth and, in fact, had already been attempting to implement such a plan between 1950 and 1970 through townwide up-zonings. Because neither the public nor the planners have real control over decision-making or the power to alter people's determination to leave the city in the first place, both groups met with only limited success.

The bi-county master plan was a classic example of technical land-use planning. Its physical ordering of space was well founded, and its recommendations, if followed by local governments, would have represented a distinct advantage over the present pattern of sprawl. Like all such proposals, however, it ignored the larger social and economic issues such as the lack of moderate-priced housing, the defense orientation of the economy, and the decline of agriculture, facing the future residents of the region. It took for granted social processes that should instead by evaluated in order to meet adequately the needs of the developing area.

Explanation and Implementation: Techniques Used in Gaining Acceptance for the Plan

EXPLAINING THE PLAN

The activities of planners are separated into two areas. Initially, they are involved in composing the plan. This is carried out in their offices without direct public input. The planners collect the necessary social information, apply the scientific principles of "good" planning to an analysis of the statistics, and discuss the technical issues involved. The composition of the plan relies on the integration of the statistical trends with the knowledge of appropriate land-use techniques. When the plan is revealed, however, the need for explaining the thinking behind it is made manifest.

In 1970, when the bi-county master plan came out, a plethora of maps, diagrams and elaborate explanations deluged the information channels to the public. The plan itself, was published in a brochure which contained elaborate four-color charts, graphs of projected growth, a variety of regional maps, and a commentary which included spiritualist-like predictions of the future society. The planners appeared at public hearings, businessmen's luncheons, and professional conferences, in some cases armed with these paraphernalis. On all these occasions, they gave roughly the same speech. Taking for granted the activities of business and government which were propelling the region on its path of future development, they followed with a sales pitch for the scientific principles of the plan. The process of explanation was usually spiced with a warning of the impending apocalypse. The people were told that only a genuine belief in the master plan and its maps could ward off the evils of their collective fate.

As might be imagined, this activity had little success. Listening to these speeches were those people who invariably were more distrustful of the label of "planner" than of the predicted land-use pattern, and the local businessmen and politicians, who knew that apocalypse or no, they would continue to make a profit under the existing institutional arrangements.

The activity of explanation is no substitute for the genuine involvement of the community in the entire process of planning, as would be the case in democratic and non-elitist planning (inductive planning). Given the constraints placed upon the bi-county board, in particular the lack of political power, explanation was one way in which to generate political support for the master plan at the local level. Despite the strong efforts of Pendleton in this area, the plan met with little success in Suffolk County. In 1971, the plan was approved by the county legislature; however, to date not a single county township has officially accepted the bi-county version of master planning, preferring instead to draw up their own plans. In Privatown, for example, a separate master plan was composed for the township by independent consultants, at a cost of $250,000 and that was adopted by the town board in April of 1975, even though the county itself had adopted the bi-county plan four years earlier. Although some of its proposals are similar to the bi-county version, such as the call for cluster zoning, the Privatown plan suggests further up-zonings of residential land to two-acre plots in order to reduce population density. Two things are worth noting about the Privatown version of the master plan. First, approval by the town board does not alter the advisory status of the planners' proposals. The township plan is still only meant as a guide to the local political leadership which retains control over the zoning power. Second, local residents will give wide support to the up-zoning recommendations, while resisting the more enlightened provisions of the plan, such as clustering. This is in keeping with their preference for limited population growth and their desire not to live in planned-unit developments. It is also probable, in view of the past use the political leadership has made of up-zonings, that the latter provision may be the only aspect of the township's plan to be implemented.

IMPLEMENTATION TECHNIQUES OF THE BI-COUNTY PLANNING BOARD

The second activity of the planners following publication of the plan involves their use and extension of the economic and political resources available to the bi-county board in an attempt to foster adherence. These techniques are sometimes called the plan's "implementation capability."

The widest used power of the bi-county board is its ability to review certain township zoning decisions. Although the local town boards have control over zoning, state law grants regional planning agencies the power to review zoning changes affecting properties within 500 feet of municipal boundaries as well as state and county highway rights of way. This power provides for the regional board's ability to override local decisions in the event that they are not compatible with the master plan or the needs of adjacent townships. Shortly after the plan was published, an attempt was made to obtain broader political powers of review. With the help of sympathetic friends in the state legislature, a bill inspired by the planning board was introduced which would have given the regional board review powers over all proposed zoning changes that affected the guidelines of the master plan. This bill, however, was defeated in 1971.

Lacking the broader powers of review, Pendleton reassessed the need for greater implementation capability and developed a strategy for dealing with local residents, party organizations, and regional planning needs. His formula was a mixture, proposing the adoption of planning guidelines for local communities, the revision of all zoning laws so as to insure uniformity of procedure in the bi-county region, and the active role of the board as an advisory body in local decision-making. Thus, the desire to challenge the principle of home rule and broaden the political powers of the board was partially laid to rest. Instead, Pendleton stressed the role of the board and its master plan as a referrent and an information source for land-use decision-making. The bi-county report on zoning emphasized this approach:

> The test of local zoning ordinances reveals that they are generally complex, confusing, incomplete, redundant and in many cases contradictory. . . . Zoning should continue to be the responsibility of

> local government; however, to insure that zoning actions do not conflict with state, local or county planning policies, existing state legislation providing for county zoning review should be strengthened.
>
> In addition, the board would continue its advisory functions including the updating of planning proposals, the dissemination of planning information, and interagency cooperation.[15]

Since 1970, under Pendleton's leadership, the board has refrained from attacking home rule directly. Instead it has attempted to exploit its advisory role. In Privatown it has publicized its opinions of all zoning proposals which threaten the master plan. In addition, the board has worked closely with developers who have attempted to construct communities using the cluster concept. Elysian builders, for example, received very strong support for their planned-unit development, Oak Forest, from the board. We shall discuss this case more fully below. Pendleton, along with the builder, negotiated with the local political leadership for support of the proposal, a number of political bargains were struck, and its was subsequently approved. In general, Pendleton's method of operation in applying the board's influence has been one of low-profile, non-partisan, behind-the-scenes negotiations. Exceptions have occurred when press statements have been released or when Pendleton has indulged in the process of explanation, such as his appearance as a speaker at public conferences.

The board's implementation capability using political review powers remains limited. Attempts at the expansion of such powers have met with sound political opposition whenever they were tried. A proposal to grant the board powers of review for shoreline development, for example, was defeated in a public referendum in 1972 held in both counties. The other area in which implementation power resides, however, that of economic review, looms as a significant category. Circular A-95 of the office of management and budget of the federal government states that regional planning authorities have economic review power over any federal funds allocated to a local area. It requires applicants for most federal grants to notify the appropriate metropolitan planning agency and give that agency an opportunity to comment on the project. Due to the private-enterprise orientation of suburban townships in Suffolk County, local governments have not yet proposed applications for federally funded projects so this economic review power has been relatively untried. In the future, however, the board may be called upon to exercise increasing control, if the need for federal funding arises to meet the growing inability of suburban governments to finance the demands of the local population through property taxes.

The Oak Forest Project: A Case Study

The planned-unit development known as Oak Forest is significant as being the second PUD in Suffolk County and a major victory of the bi-county board. As negotiations developed between the elitist sponsors of the plan and the residents living in the area to be most affected by the project, Oak Forest is also significant as an example of the way the town board operates as an apparent pluralistic mediator between developers of housing and contending residential groups. Planner recommendations allied with the housing developer proposals were tempered by the political negotiation process touched off by the announcement of the plan. This example of "successful" planning, however, left the local residents who were dwelling in the vicinity of the proposed development in total opposition to its presence.

The notion of cluster development is a planner's answer to the chaos of suburban sprawl. It was first used successfully by the developers of Radburn, New Jersey, in 1929.[16] In the place of individual ownership of housing lots, which eventually supports the pattern of sprawl, construction is clustered on the same amount of land which would be used if ordinary development were to proceed. Placing homes closer together liberates a large area of open space which then becomes community property. The planned-unit development has two important advantages arising from this structural innovation. First, it preserves open space within the community by maintaining a planned public space to housing ratio. This is absent in single-family developments, which privatize open space for use as backyards and front lawns. Second, the proximity of housing and the clustered arrangement of developed land carries with it great benefits to the builder in the form of lower infrastructure costs. The builder has fewer roads to pave, fewer utility lines to install and more efficient use of material and labor.

Despite these benefits to residents and builders, the PUD concept has not caught on with the suburban public. The fear of losing open space to even further development and, therefore, denser community growth is one reason for this reluctance to buy the planners' idea. A second reason is the public's preference for ownership of an individual plot of land and the distrust of sharing community resources with other residents. The privatism characteristic of suburban life works against acceptance of the idea and thereby supports the continuation of spread city.

In 1971, a local developer in Privatown filed a petition for a 900-acre PUD that would contain a population density somewhat lower than existing zoning requirements. It would contain a large golf course, a common open area, schools, gas stations, and an area for light industry.

At the time, it was also revealed that this PUD coincided with a master plan proposal for a new center of population to be developed in the same region, and, furthermore, that the selection of the site by the master plan was influenced by Elysian Homes' large land holdings in that area. The publicity released on this PUD indicated that the bi-county planning board had worked for three years with the developer, Burke, owner of Elysian, and the local town officials in order to formulate the planned development of the community.[17] This was corroborated by a member of the local town planning board who is also a resident of the area. He stated in an interview that any development of this magnitude would as a matter of course be cleared first with the local planning board. Instead, he heard about it only when the rest of the public did, at that time the petition was filed. Cooperation between bi-county, the developer, and the local political leadership on developing this plan, and in keeping it secret from the public because PUDs are a volatile issue, is strongly indicated.

Once the plan was announced, however, the anticipated furor erupted among the local residents. Located in the south-central part of the island, the region involved is a relatively poor one. The two major communities in the area consist of elderly people and minority groups. Although suburban development had reached the region, local residents were still antagonistic to further growth and expressed specific needs which they felt were being ignored by the Privatown government. The planning board took on the task of meeting with the local residents to hear objections to the proposal. These hearings were called "information meetings" as the plan was explained to the residents. In the months which followed, it became clear that no input existed between the local planning board who ran the information meetings and the actual sponsors of the project. The hearings were set up to "sell" the plan to the residents without arrangements to include their substantive involvement. These meetings served as a safety-valve measure to mollify resident unrest. In the stop-sign controversy, discussed below, the citizens of Privatown encountered the same use of the local planning board. They too expressed the observation that it existed merely to siphon off public discontent and to project the facade of professional planning.

By the time the first town board public hearing was held on the zoning changes necessary for the development, the fait-accompli decision-making process of the planners, politicians, and developers was apparent to the local residents. It was announced that the developer, Burke, had negotiated for four years with the local town officials on the nature of the project, and that the plan had been worked out between them. The Oak Forest development was to contain a number of features which it was felt would be attractive to the local residents. It involved a lower population density than ordinary residential zoning, thereby controlling population growth. In addition, the homes would all be moderate- to high-priced, and no low-income housing would be built.

At the first and five subsequent public hearings, vociferous objections to the PUD were made by local civic associations and the local school boards.[18] A separate ad hoc association was also formed in the area called the United Moriches Taxpayers Association. Its expressed intent was to block the plan by acting as a politically motivated interest group. One of the associations spokesmen, who represented the elderly residents, stated that it would take ten years before the development would be generating enough taxes to support the demand its residents would make on local municipal services.

The antagonism expressed by the people against the plan ignored the possibility that development would proceed in any event with or without clustering. Local residents, it appeared, were willing to take their chances with the ordinary operation of the self-governing society than with the proposed plan. The support of additional schools and municipal services by property and other forms of taxation were their immediate concerns. They felt that any such development proposed and supported in part by the planners and local officials would serve to make their region more attractive to people and thereby hasten its suburban growth.

At this stage, the local school boards responded to the apparent inevitability of the project's approval by the town board. They entered into private negotiations with the developer to obtain resources for the support of the school districts. In an interview with a local school official it was revealed that the obvious political power behind the proposal made the private negotiations imperative in order to extract some concessions from the developer before the actual decision was formally announced. This early bargaining was inconclusive as each side adhered to a position that the other's proposal was unacceptable.

The final group to enter into the public debate on the Oak Forest project was the representatives of the minority residents of a large black ghetto located in proximity to the proposed site. This group was joined by civil rights advocates, who had been operating across the country. They proposed that low-income housing be included in the project in order to meet the needs of the less affluent residents of the region. Appearing in large numbers at all public hearings, their collective voice, nevertheless, remained an outside influence on the substantive negotiations, because advocates of low-income housing faced a major and commonly understood task in Suffolk County. Only considerable political muscle could hope to overcome the local resistance to subsidized housing—the anathema of suburbia.

The public hearings were a community stage upon which apparent pluralist bargaining was being carried out. Behind this facade, however, political, social, and economic realities had already combined to create formidable constraints on the public's participation in the plan. Backroom negotiations between the bi-county planners, the developer, and the local politicians set the tenor for the public hearings. Community residents were faced with an elitist proposal and were aware that only their political veto power could create a negotiable situation in what was otherwise a fait accompli. The relative positions of the interest groups responding to the proposal were as follows: Practically all community residents opposed the plan as encouraging growth. The ad hoc taxpayer groups and civic associations represented the sentiments of the majority. Only two groups, the school boards and the civil rights advocates, suggested ways in which the developer could alter his plan to fit their needs. The ghetto representatives called for the inclusion of subsidized housing, but their demand was ignored because of the political realities of suburban life. The school boards meeting privately with the developer requested to be bought off by being provided with land and monetary compensation to partially offset the costs of providing additional facilities. Among all the community residents participating in the public hearings, the latter group was the only one to engage in actual bargaining encounters with the developer. Ironically, it might be pointed out that the elderly residents did not identify with the need for subsidized housing despite the obvious use by this age group of such arrangements in other areas of the country. A possible coalition between the old and the minority poor did not materialize, nor was it ever considered.

The town board's eventual decision on Tuesday, December 19, 1972, was notable for its tight revision of the plan as well as the verification of its inevitability. Although it was approved, drastic revisions were made in order for the PUD to conform more closely to what was taken by the political leadership to be the needs of the local area. As a matter of course, the board approving a re-zoning can attach restrictions, a "covenant," to the developers right of land-use. While this is done in many cases, the scope and nature of the restrictions placed on the project were an unprecedented response of the Privatown political leadership. Never before had the board taken an elitist plan and made it conform with such tight control and well thought-out planning ideas to the perceived needs of the local area. It is the best a community can hope for, given the present-day planning process.

Councilman Hymer, often the spokesman for the board, issued the following statement: "We have placed some very strict controls on this project in order to minimize the objections of some residents in the community." He said that he realized it might be "politically expedient" to turn down the project without any thought of future consequences, "but because of the far-reaching nature and the complex questions involved, we have revised the applicant's proposal in the best interest of the residents of the community."[19]

Among the revisions were (1) the reduction even further of the maximum housing units to 1,400 from 1,632, (2) of all apartments in the PUD 80 percent could be no more than a one-bedroom unit, (3) the three gas station re-zonings were denied, (4) the developer had to covenant that the golf course area "forever remain" as a recreational facility, and in no event might the land be used in any other category, (5) 200 acres of additional open space would be deeded to the town at no cost (a tax-deductible contribution for the developer) for park land, (6) finally, the developer agreed not only to donate land for schools but in addition to erect school rooms on the site of the development to the extent that existing school facilities are unable to accommodate the additional school children of the PUD. These buildings would then be leased to the appropriate district, if desired, for a period of up to five years, at an annual rate of $3 a square foot.

Oak Forest is an example of successful planning by the bi-county board in that the developer and the local public officials cooperated to create a settlement pattern which was called for by the master plan. Yet, these "good plans" were quite contrary to the wishes of the majority of the people, for "good planning" is itself a response to only a small segment of the public, largely planners themselves (see Figure 4).

Forming a bridge between the discontent of the local residents and the wish to exclude housing for the poor and the elderly, the town board responded with restrictions. Operating according to what it perceived the local needs to be, it altered the elitist plan to eliminate some of the features that were objectionable from the viewpoint of the local residents. The feeling of powerlessness, while not dissipated, was, nevertheless, assuaged to some degree. The elitist mode of decision-making, however, was not changed any more than was the physical approach.

The negotiations on Oak Forest involved important influential groups

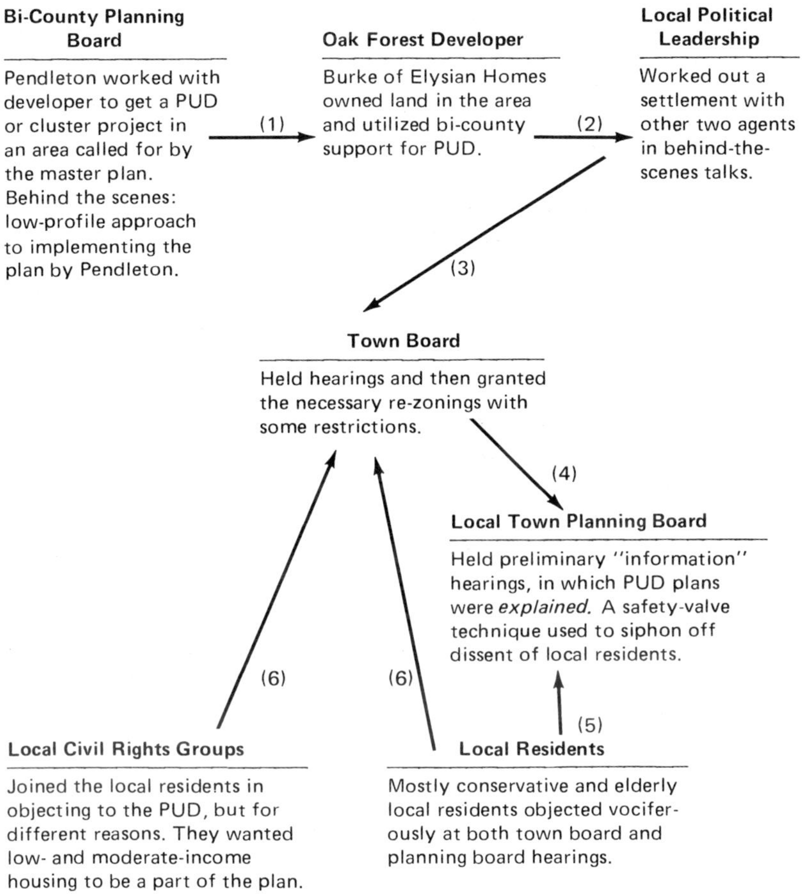

Figure 4: THE OAK FOREST CASE STUDY: NEGOTIATIONS

in Privatown. School board, bi-county planners, developers, local civic rights associations, the local planning board, and ad hoc community groups participated in the process which culminated in the town board decision. A critical point is that the participation in the bargaining process was not at the same level of interaction for each of these groups. The public was essentially excluded from the initial conception of the plan. Its real needs, the needs of the elderly, the poor, and the minority group residents, were ignored. In this type of decision-making, which is characteristic of the wider society, the public was allowed to participate only because they could threaten to exercise a political veto vote against the local politicians at the polls. In order to obviate the political potential of such a response pseudo-forums were set up. These "information" meetings and public hearings gave the appearance of open negotiations despite the actual conditions of elitist control. In the end, the local political leaders demonstrated the response of "good government." They tempered the planners' and developers' proposal in ways which gave the appearance that they were being responsive to the public. Given the fact that the developer owned the land in the first place and that the planners had designed the proposal to fit into the structural view of what constituted the long-term needs of the entire region, the representatives of the polity could easily have defended their decision against the persisting community discontent.

This decision-making process involves a hierarchy of power and an order of participation. It is bargaining and negotiation of a very special kind. It represents a limited form of planning as well as a limited exercise in political responsibility.

The Suffolk Community Development Corporation

The bi-county master plan indicated that in order to meet the needs of the projected population, 283,000 housing starts were needed by 1985.[20] Of these, the plan said, roughly 18 percent would have to be publicly assisted units. Recognizing the fact that the board had no way of implementing such a goal, it appeared that some alternative agency might have to pursue the task of getting adequate housing for the economically deprived groups.

In 1969 a coalition of civil rights groups in consultation with the bi-county board and the Suffolk County Department of Social Services, organized a drafting committee to look into the development of other institutional forms which might be able to urge the construction of adequate low-income housing. Among participating groups were the Smithhaven Ministries, an interdenominational religious organization, the Human Relations Committee of Suffolk County, the Economic Opportunity Council of Suffolk, and the Long Island Council of Churches.

The proposal recommended the creation of a nonprofit housing corporation that would serve as a link between those community organizations that were attempting to bring about the construction of adequate housing, and the government programs, and private business interests which operated in the county. A major impetus to the creation of such a semiprivate agency was the existence of the statewide Urban Development Corporation which was given the power, at the time, to override local zoning in order to pursue the goal of construction of subsidized housing.

> The Drafting Committee proposes the organization of a corporation, Suffolk Community Development Corporation, to include representation from business, labor, civil rights groups, education, government, religion, and community organizations, as essential toward meeting the overall needs and goals for public and publicly-assisted housing indicated by the Bi-County Planning Board. The combined support and resources of all these sectors would make it possible for such a corporation to undertake systematic, large-scale, highly visible efforts both on its own and as an educator, catalyst, and resource for both public and private action toward meeting the housing needs of all persons in Suffolk County.[21]

The key expression in this declaration of intent is "highly visible." We have already indicated that the explicit goals of land-use in the master plan were for the most part ignored due to bi-county's lack of political power. The leadership of that agency thus had to engage in behind-the-scenes, low-profile maneuvering and exert influence as a special non-partisan interest group in order to pursue its land-use goals. An explicit social goal, such as adequate housing, was therefore infinitely harder to pursue, and any identification with such a measure would certainly jeopardize the board's credibility in its negotiations with political and economic interests. Whatever hope of realization of such a goal would have to come from another more visible agency linked only indirectly to the board. The SCDC was the institutionalized answer. In late 1969, the drafting committee chose a director, Philip Burns, for the new organization. Burns had come from a Connecticut office of city planning where he had enjoyed a measure of success in administering urban renewal programs. Burns had achieved a reputation of being able to deal with the "power structure" on the question of subsidized housing. In the words of one member of the drafting committee, Burns was picked for the position because he stated in his interview, after being told of the bleak outlook for the negotiated pursuit of social goals, that dealing with such a firmly entrenched, private enterprise-oriented power structure would be "a real challenge."

SCDC differs substantially from its statewide big sister, the Urban Development Corporation. The latter was set up by an act of the state legislature, in April 1968, with mandatory powers over land-use policies throughout the state. UDC, thus, had the political clout to implement its plans, as it had been given the power to override local zoning. Subsequently, this power was removed for reasons which are similar to the restrictions placed on SCDC.[22]

The Suffolk Community Development Corporation utilized the powers of UDC in a peripheral way. The local animosity generated by the UDC required SCDC to underplay its connection with it. SCDC's main task was to cooperate with the other more powerful political agents in order to work for the construction of subsidized housing. Economically and politically it relied almost exclusively on existing federal legislation and its ability to convince local entrepreneurs and investors to take advantage of the programs offered. Despite this liaison role, SCDC did receive its operating budget from UDC and shared the UDC power to obtain loans at the regular market rate. Burns has stated in an interview that his method of operation as head of SCDC involved two basic procedures, that of working with broad coalitions, and that of demonstration projects.

PLANNING THROUGH BROAD COALITIONS

In a somewhat difficult situation, SCDC attempts to work with broad coalitions of people who then carry on a political struggle in their community to "try to educate, coerce, or cajole the municipality into cooperation with the proposed project."[23] At the very most, a coalition probably only controls some of the land necessary for a project, hence it must actively negotiate with an often hostile community to agree to its efforts.

In this type of situation Burns tries to keep out of politics entirely and prefers to work closely with the coalition itself. One reason for this is that SCDC is stigmatized by the local fear of the "imperialistic" powers of UDC. Another reason for this is that in order to operate at all, SCDC has had to make an unwritten agreement with local leaders not to interfere in local politics. This differentiation of function from bi-county enables the latter to pursue political negotiations with local leaders over the broader goals of the master plan, while it allows SCDC to cooperate openly with anyone who might otherwise bring it under political criticism, provided that that individual is receptive to a low-income housing project. Thus, bi-county retains its low profile to pursue its broader program, and SCDC is freed to assume a highly visible profile to pursue its more specific aims. Recently, for example, SCDC openly cooperated with town boss Middleton on the creation of a low-income housing project in Privatown. In contrast, bi-county could *not* have cooperated in the same open manner with the Republican party boss and still retained a nonpartisan image.

An example of work with a broad local coalition, although not local to Privatown, is that of the Greenport redevelopment project. In this case a group of business leaders in Southold township were working on a project to revitalize the town and its declining economy. Part of the businessmen's concern was a large downtown area of substandard housing that had to be eliminated before any revitalization could occur. Burns, in his role as a low-income project entrepreneur, received word of the businessmen's predicament. He approached the coalition and sold them on the idea of developing a plan which would rehabilitate the area of urban blight as well as build a new commercial center. Rather than just wipe out the housing and relocate the indigent migrant farm workers, many of whom were black, Burns persuaded the coalition to build low-income housing and use the tax shelters afforded by such a project to support other revitalization construction efforts. He and his staff then drew up a master plan for the area with a financing scheme that would deal with both the economic and housing aspects. This was accepted by the business community.

In describing his attitude as a planner, Burns emphasizes that he is a pragmatist willing to work with anyone. If the possibility exists that low-income housing can be built in a specific area, SCDC is willing to pay the local politicians' price for favorable behavior from the town board. Usually in the case of SCDC-type projects, the price is that he work with a developer or construction firm associated with the local political leader.

DEMONSTRATION PROJECTS

Typically, SCDC is invited by an entire community to work on a project that will help satisfy its housing needs. SCDC supplies a service that helps with planning architecture, engineering, and financing. In an interview Burns stated that this is the ideal situation for a nonelitist planner to work in, because the entire community can be involved in the project. In this type of situation Burns uses a mixture of advocacy and democratic planning techniques. Sometimes he leads the community by suggesting what he considers to be viable goals, a procedure used in advocacy planning, and at other times he allows the entire community to decide the goals of the project. In opposition to the elitist type of planning practiced by most planning boards, these techniques utilize continuous feedback from the community in making decisions and are, therefore, more democratic. Burns employs these democratic techniques to establish a dialogue in the community over the nature of the plan. A brief summary of the Wyandanch Project case study can serve to illustrate the method of approach used in demonstration projects. Wyandanch is a predominantly

(85 percent) black community in the Township of Babylon on Long Island. Burns was invited by this community and asked to help develop a project which would revitalize the housing in the area.

His procedure was to define the situation in such a way that people would discuss explicitly what they wanted. He continually asked them to formulate an answer to the question "What is it *you* are after?" He thus solicited and received feedback on the step-by-step planning process of the project. Burns did not want the plan to be swallowed up by the existing municipal government as so often happens in community planning. He created a situation where the master plan of the project became the center for a community dialogue.

Erroneously labeled as a low-income project by the surrounding communities, the actual breakdown of housing in the final plans called for 70 percent moderate income ($5,000 to $9,000), 20 percent low income (below $5,000), and 10 percent senior citizens. The apartments would be built in twenty-nine two-story buildings and each would have a separate entrance. The twelve-acre site would also be clustered so that open space would be preserved. The project's master plan, thus, had elements in it which are characteristic of the best type of planning. Not surprisingly, however, opposition did emerge in the adjacent, predominantly white, areas.

It is hard to imagine decision-making constantly being carried out at mass meetings. As a means of overcoming this drawback to democratic planning the community agreed to form a representative body, an executive board, independent of the municipal government, which would meet with Burns and make decisions. The executive board, however, was supposed to be self-reliant and independent of SCDC. The members were required to return to the people to ask for their approval and suggestions as the project developed.

Once this machinery was in operation, Burns then attempted to locate a developer who would cooperate with the plan. This was done along very specific lines. Subsidized housing is financed under HUD 236 legislation. Within this framework, however, a developer can only realize 5 to 10 percent profit. This is much too low for most developers who enjoy returns of as high as 20 percent. The way SCDC entices developers is through the opportunity of acquiring equity for tax shelter purposes and not for the realization of profits. Under the total package of existing legislation a developer can receive a loan from the government which covers as much as 90 percent of the costs of the project. Then, by using the entire amount as equity, he can write off substantial depreciations to be balanced against other earnings. As the following suggests, the returns under this arrangement are substantial:

> With the actual cash equity of only $33,000, less than 3%, a not unusual proportion, the developer of a $1 million project can get a cash return in less than two years equal to five times his investment.[24]

Ironically, the exact mechanism which is used to entice construction of the project is now contributing to its stalemate before completion. Wyandanch has been famous for the resistance of neighboring white communities such as Deer Park. Formidable opposition also arose within Wyandanch itself. The executive board of the community, having worked on the project for two years, has come to understand how the tax shelter works and they now feel that they are entitled to some of the equity which the development represents. From an initial honeymoon period in which Burns was viewed as a white folk hero, the situation changed to one in which he fell into ill repute as a handmaiden of white capitalist interests. The decision-makers themselves, as we have already seen, have been successfully tempted by the opportunities for personal gain presented to them. The project was first held up because the executive board demanded that part of the equity which was used to attract the developer to build the project be given to the board as payment for their efforts. The "concerned citizen" orientation of the board has been reduced to that of the private, self-interested citizen.[25]

The Wyandanch project eventually failed. However, internal bickering was not the cause. Despite Wyandanch's modest size, the attacks from the neighboring communities increased. Burns violated his own method of operation in a somewhat risky attempt to counter this political pressure. On October 1, 1972, the State Urban Development Corporation announced that it would be ready to override *any local zoning* restrictions in order to build the project. They felt justified in such action, if needed, because the local community had requested the project. Militant activity on the part of the surrounding areas could jeopardize it because ultimately the Babylon town board had to rule on the necessary re-zonings. At the time of the announcement, Burns said, "I think this is going through regardless. We are ready to ride it out if there is a lot of flak. It would be our first project on Long Island—and maybe our last."[26]

Threatened by political pressures from outside the community but within the town board's jurisdiction, Burns had no choice but to counteract by using extra-local political power. If outside areas would be allowed to object to what a local community adopts as a planning strategy, then the method of demonstration projects would be useless, and SCDC would cease to exist.

As a result Burns fould himself in a pincer move. On the one hand he faced community pressure over his control of equity, and their cooperation temporarily ceased. On the other hand, he had to fight a political battle in order to preserve the integrity of a community-supported demonstration project.

The threat to use UDC's powers in the Wyandanch case, however, backfired. On May 22, 1973, an announcement was made that a Babylon citizen's group had blocked the final completion of the Wyandanch project and was lobbying in the state legislature to restrict UDC's powers of zoning only to the central city. The power of the white middle-class Republican majority on the island prevailed, and in June 1973, the New York state legislature agreed to strip UDC's power to override local zoning in suburban areas. UDC retained this significant power only in the case of central city projects. The Wyandanch project was defeated by the township board's vote. UDC and SCDC lost at the political game in a most crippling way. In a dramatic move, nine days after Governor Rockefeller signed the bill stripping UDC of its suburban powers, Burns announced his resignation from SCDC. In addition, he stated he would become director of project planning for SPT Enterprises, a local Privatown-based private developer. Burns thus joins David Seymour and the private business associates of Middleton. Entrepreneur Seymour, along with his firm SPT, has been the only successful builder of subsidized housing in Privatown, the Pleasant Village project. Burns, in his role as planner interested in the development of moderate and subsidized housing, joined the business community because this was the only area in which the realization of such housing had occurred. Although a search for a new director of SCDC was announced, the ability of the agency to function without UDC's powers of implementation is in grave doubt.[27]

Conclusion

The Nassau-Suffolk planning board is the public agency of regional planning on Long Island. The Suffolk Community Development Corporation is a semi-private agency which attempts to plan for the social goals of suburban growth, in particular for moderate-priced housing. Both appear to labor under virtual political impotency with limited resident support. The planning process, as it is usually practiced in the society, makes planners advisory bystanders to decisions that are being carried out elsewhere—by political leaders and private businessmen. They do the best they can with the low prestige of an advisory role. Initially, this status is reflected in the bi-county planner reliance on technical rationality and physical design principles. Rather than incorporating the explicit pursuit of social goals, such as the construction of communities with a mixture of income and racial groups, their approach aims to appear politically neutral. The latter needs are pursued by a select number of planners, like SCDC, called advocacy planners, who help fight the political battles of the socially disadvantaged. By confining their approach to the manipulation of the physical form of growth, the bi-county planners are using the least offensive technique of control to the residents and the political leadership of the region. As we have seen, however, even this effort has had little success, although the design principles involved in the master plan comprise an excellent minimal attempt at controlling suburban sprawl.

Second, advisory status creates the need on the part of planners to become politicized in behind-the-scenes bargaining and press for their plan's acceptance by negotiating with builders and local political leaders. The desire to strike up political bargains with the local township political leadership gives bi-county a distinct nonpartisan character, because the planners are willing to work with the party organizations of the local townships regardless of political affiliation. In this alliance the bi-county planners are as much a victim of their political negotiating position as of their elitist conception of planning procedures. Because the planners must bargain with the most powerful interests in the township, they are often identified by the public with those same interests and, therefore, distrusted. In the case of SCDC head Burns such elitist identification became complete when he quit the agency and joined the private firm building in the township of Privatown.

The experience of bi-county and SCDC reveals that planners' ideas, whether physical or social, find little support among the majority of white middle-class suburban residents. Despite full local community support in the SCDC demonstration project for Wyandanch, for example, the local township board opposed anything but the low-density suburban mode of home construction for the racially segregated area. In Suffolk County local home rule and community control by the local political organization supported by the white middle-class are effectively in charge of the decision-making process which could potentially achieve viable social planning. One important implication of this experience is that social change in the submetropolitan region may be more difficult than often anticipated. With community control working in the townships to preserve property values, while allowing the local political leadership to support themselves and their party through control over land-use, attempts at planning for the alleviation of the housing inequities and the sprawl pattern of growth, using those same controls, may be extremely difficult. Those planners and individuals wishing to challenge the direction of social growth are, in effect, making a *political* challenge, because they disagree with the social values embodied in the present land-use pattern. Consequently, they are shunted into the larger pluralist arena of political bargaining within which all other interests appear. At least at the local level, this state of affairs enables those in control of land-use decisions to retain it, while effectively tying up all other interests in the difficult process of pluralist negotiations.

As we have seen in our discussion of exclusionary zoning, it may be easier, in the light of the above, for local residents interested in planning to go directly to the courts and challenge the status quo in a judicial rather than political way. Recently, in Suffolk County, there has been increasing activity in this direction. Lacking genuine support or resource controls, however, attempts at planning will no doubt continue to achieve only limited success in the region. Perhaps the only hope for broad support of planning involves the confluence of resident desires to maintain property values and low population density with the planner and conservationist wishes to preserve environmental quality. In such case a three-way alliance is possible, and planners may yet achieve some rational control over land-use if, at the same time, they also agree to measures which would restrict population growth of the region, especially of the low-income and non-white segments of the population. Rather than being a political interest of a small group of "eco-freaks," ecology may become the rallying cry of the relatively more powerful elements of the Suffolk County population, in particular the white residents, planners, and conservationists, in a last attempt at controlling the process of suburban growth.

NOTES

1. *Newsday*, 5/6/66.

2. This is also close to the United Census occupational definition of planner. Recent evidence suggests that current graduates of planning schools are working less for local planning commissions producing master plans and more for universities, private planning firms, and state and federal government agencies. The skills emphasized by the current planners, however, have remained the same—physical approaches to

land-use and housing. Cf., Donald A. Schon et al., "Planners in Transition," *AIP Journal,* (April 1976), pp. 193-203.

3. Alan Altshuler, *The City Planning Process,* Ithaca, N.Y.: Cornell Univ. Press, 1965, p. 3. This text is a good discussion of the limitations of municipal planning.

4. This approach sharply contrasts with the social role of planners in socialist or social democratic countries, such as in Northern Europe. Land-use, along with social and economic activities, are integrated into overall growth schemes which are often tied directly to political parties holding office. In Norway, for example, local elections are run on the basis of proposed plans rather than competing personalities. Holding office, then, depends upon the officials' ability to implement the integrated master plan.

5. One issue which failed in 1972 to achieve public support was the proposal of the bi-county Board to assume zoning control of the shoreline.

6. See Gans' reply to Jane Jacobs in Herbert J. Gans, *People and Plans,* New York: Basic Books, 1968, pp. 25-33. As noted in footnote 2, planners are still taught these skills.

7. Altshuler, op. cit., p. 354.

8. Paul Diesing, *Reason in Society,* Urbana: Univ. of Illinois Press, 1962. Both Max Weber and Karl Mannheim explored earlier aspects of this view.

9. In the 1960s, planners attempted to move beyond the elitist approach. This was evident in an extensive and ongoing public debate, some examples of which are: Paul Davidoff, "Advocacy and Pluralism in Planning," *AIP Journal,* (November 1965); William Petersen, "On Some Meanings of 'Planning,' " *AIP Journal,* (May 1966); Ernest Erber (ed.), *Urban Planning in Transition,* New York: Grossman, 1970; Robert Goodman, *After the Planners,* New York: Simon & Schuster, 1971; Murray Bookchin, *The Limits of the City,* New York: Harper & Row, 1974.

10. J. Bellish and M. Hausknecht (eds.), *Urban Renewal: People, Politics and Planning,* Garden City, N.Y.: Doubleday Anchor, 1967; Robert Caso, *The Power Broker,* New York; Random House, 1975.

11. *Newsday,* 4/29/65.

12. One social function, therefore, of master plans is often to provide speculators with the codification of future demographic and regional developmental trends. The social functions of land-use controls in general have been explored in Chapter 4.

13. Tom Morris, *Newsday,* 7/13/70.

14. Nassau-Suffolk Bi-County Planning Board, Master Plan, 1970.

15. Bi-County Planning Board, *Land-Use Plan,* 1970.

16. Without discussing the many limitations of the Ebenezer Howard school of town planning, the "garden city" approach is presented by Clarence Stein, *Towards New Towns for America,* Cambridge, Mass.: MIT Press, 1971.

17. *Newsday,* 11/10/71, p. 6.

18. The area of the proposed PUD is one of low to moderate socioeconomic status, and, thus, the concern of the residents was real. The Shirley-Mastic area, in fact, is the poorest in Privatown. Over 18 percent of the people residing there are sixty years or older. For a separate study of the same region, cf., A. Bruce Borass, unpublished M.A. thesis, *Resistance to Urbanizing Influences,* SUNY at Stony Brook, (December 1973).

19. *Newsday,* 12/21/72, p. 1.

20. *Master Plan,* op. cit.

21. *SCDC Prospectus*, 1968, p. 11.

22. In addition, UDC defaulted on loans in February 1975, causing a financial reaction that eventually engulfed New York City as well, and causing the state to retrench and reorganize its housing program. To date, the ripples of this planning superagency failure are still being felt in the region.

23. Statement made by Burns in an interview.

24. Long Island *Press*, 12/9/71.

25. This is another example of the social problems of the control of decision-making. In this case it is a group of private citizens and not the town board. Several community development corporations in the central city, such as in ghetto areas of Brooklyn, N.Y., have also experienced this tendency for political appropriation and scandal to become limitations on implementing plans.

26. Burns, op. cit.

27. *Newsday*, 6/16/73.

Chapter 7

THE LOCAL POLITY AND THE SOCIAL ADJUSTMENT PROCESS: A Case Study of the Vulnerability of Party Organization

It is helpful to regard the planning process in Privatown as an allocative system of rewards and costs. Land-use decisions can then be analyzed from this perspective. For example, successful petitioners for changes of zone receive locational economic advantages for their enterprises. The local political organization also benefits from such decisions, as we have seen, by exercising its control over the town board. The effects of such decisions on the public, however, can vary considerably. On the one hand, favorable zoning for business can enhance the development of the township and result in lower or more stable property tax rates. In that case the public receives the benefits of economic growth and also enjoys the longer term multiplier effects of greater regional prosperity. In Privatown, whenever the local political organization, entrepreneurs, and the public have anticipated such gains, there has been concordance in the zoning-decision process. The rewards have been highly visible, and potential public costs which might arise in the future, such as increased pollution, have remained hidden.

On the other hand, many proposals are of benefit to local business but bring with them direct costs to the local residents, such as increased traffic hazards. In such cases town board members are required to weigh public disadvantages against private gain. They do so by hearing planning board recommendations and by scheduling public hearings where the voices of residents affected by development can be heard. At times, business interests prevail in the decision-making process despite such a procedure. The public interest expressed at such meetings can be totally ignored and the local residents can be saddled with the unpleasant effects of growth. As we have seen in previous chapters, these decisions are very unpopular and often represent a large financial gain for the political leadership, which then assumes the risks of resident displeasure. At other times, however, the town board may be very responsive to citizen interests and take them into consideration in a variety of ways. At one extreme the board may respond to discontent by rejecting the plan and refusing the re-zoning. In most cases, however, a covenant or rider is attached to the proposal, which is then approved as a package agreement.[1] The covenant may redistribute private advantage by tailoring a proposal to conform closer to public needs. Such conditions may take many different forms, from requiring builders to increase the number of fire walls in a development's housing to increasing the amount of open space within a project for recreational purposes. The town board enjoys this wide latitude in its decision-making and draws upon several areas of interest including that of professional planners before final deliberation.

There are at least three general features of this township planning process which are important for the following case study. The primary characteristic of land-use in Privatown is that zoning decisions are initiated by private interests. In appearance the local political organization plays the role of mediator between businessmen and local residents, but also participates along with entrepreneurs in behind-the-scenes transactions. The polity, however, is not directly involved in the planning process and reacts to it as development patterns emerge from the landscape. The local residents join the planners as monitors of township growth with limited powers. The substantive proposals and deals which form the base for suburban development are removed from public discourse in the preliminary planning stages.

Second, the benefits of growth to businessmen, when visible, are then potentially open to renegotiation as local residents become aware of proposed developments and react to them. Depending upon the strength of such a response, considerable renegotiation may result. The town board as mediator facilitates this stage and may alter re-zoning proposals or reject them. Such renegotiations are, however, based upon perceived benefits and costs that do not always reflect reality. This is especially the case when long term hidden costs of growth are involved, because these often remain outside the purview of even the professional planners.

Third, the parties involved in the planning process, with the exception of the professional planners, take a very limited view of regional needs. Most residents are content with the pattern of growth and have little interest in increasing the monitoring role of the polity or in acquiring greater participation in the planning process. The "limited liability" of resident concerns is also reflected in public support for the ideology of the self-governing society. It is generally felt that the destiny of the region lies in the continued health of the business community and in the town board's continued efforts at fiscal zoning to preserve property values. There is, therefore, a lack of support for broader visions of a planned future and greater public participation in choosing the goals of growth so that private interests seem to prevail. Challenges to the political organization after a land-use decision are rarely made because they involve legal action. Residents can go to court to reverse a re-zoning decision or sue to change local government land-use politics, but few public protests are ever carried this far. Such action requires an organized collective effort by community residents and supportive financial resources—two properties local citizen groups most often lack. Local party control, therefore, is not very vulnerable once decisions are made.

The Hallock Road Stop-Sign Case Study

On May 16, 1967, the parcel of land making up the intersection of Hallock Road and the recently completed "limited-access" Nesconset Highway was down-zoned for a commercial shopping center. This was done despite the objections of the local residents and with a denial recommendation by the township planning board. This intersection is in the area of a large single-family housing development, and, in addition, the change would increase the strip zoning along the highway. On a previous day, the planning board said of this re-zoning that it unanimously recommended

> denial of this change of zone because many large shopping center sites have already been approved to serve the area but have not as yet been developed. The master plan shows quite clearly the two major commercial areas in this part of the town. The board also feels that such a change *could create traffic problems on the by-pass.* (Emphasis added.)[2]

The down-zoning was only one of a series of discordant rulings affecting Nesconset Highway 347. As early as 1962, the bi-county planning board sent the following letter to the Privatown town board:

> Gentlemen:
>
> In recent years the Planning Board has on several occasions notified the Town Board of their policy regarding the Nesconset-Port Jefferson Highway. . . . The Planning Board has consistently taken the position that the right of way was acquired . . . for the purpose of taking the through traffic away from the villages. If we consider applications for rezoning along this highway, then we will, in effect, create new villages and business areas and defeat the purpose of this road.[3]

This recommendation was subsequently ignored, and the strip zoning of the highway continues to the present day.

In 1967, families began moving into the newly completed Strathmore development. At that time, the surrounding land was undeveloped with some agricultural use, and Hallock Road was unpaved. As is the case in a rapidly developing area, none of the new suburban residents had any objective signs to indicate what was to follow. They were also unaware of the 1962 down-zoning.

In the early part of 1969, it became apparent that the corner lot was going to be developed. By then, the planning board decided to hold public hearings in the area because a Valuehaven shopping center was going to go up on the lot, and a measure of the potential effects of the development was required as a matter of procedure. A local resident, who subsequently became involved in the political controversy over traffic controls, remarked that at the time of the township planning board hearing, most residents believed that their interests were being considered in the planning process:

> At the time [1969] I still believe that the planning board had some kind of power to decide issues or, at least, influence them and so we went all out at this planning board hearing only to find out later that its main purpose was to give people the opportunity to let off steam without the possibility of having any effects whatsoever since it is the Town Board that makes all the decisions.[4]

In this case it was only after the hearings that the nature of the planning board meetings suggested itself to the residents. They came to realize that the true process of decision-making took place elsewhere beyond public scrutiny and that planning board meetings were actually safety-valve pseudo-forums staged by the local political organization without meaningful public participation.

The residents appeared at this early meeting because they were disturbed by the possibility of an increase in the traffic along Hallock Road. In 1968, just prior to the beginning of construction, the road itself was paved and they could see that it was a bit wider than the average suburban street in their development. As one resident indicated:

> The people of Hallock Road proper got together informally and had a brain storming session where they thought of all the possible techniques that could be used to hold down the volume of traffic. Many measures were considered and subsequently proposed including stop signs.[5]

The measure that was most favored by the residents, however, was the actual severance of the road. Their reasoning for this was based upon the self-evident fact that the street was part of a residential community, and any planning for the development should have made all roads within it residential. Hallock Road, thus, appeared to them as an aberration in an otherwise residential area. The oversight in planning which it signified could be corrected by converting it into a road consistent with the others. This was why severance seemed the only genuine proposal to make.

At that time, the residents also considered attempting to block the building of the shopping center, but since the down-zoning was already passed, this appeared to be a futile gesture. They were also threatened by the building on the other end of the road of a large, multi-acre parking lot for the State University of New York at Stony Brook. Residents in that area of the project, called the "M" section, were disturbed by the height of the parking lights in the lot, and they were interested in petitioning the town board to have them lowered. The people in the "M" section, however, objected vehemently to a severance of Hallock Road as they felt this would prevent access to their section by emergency vehicles. See Figure 5.

The Hallock Road residents wanted the support of the "M" section in dealing with their traffic problem. They called an informal meeting to discuss a unified community strategy, and eventually agreed not to single out a demand for severance, but, instead, to present all constructive possibilities to the board with a request that the traffic hazard should be eliminated. They also worked out an agreement by which the Hallock residents signed an "M" section sponsored petition to lower the SUNY lot

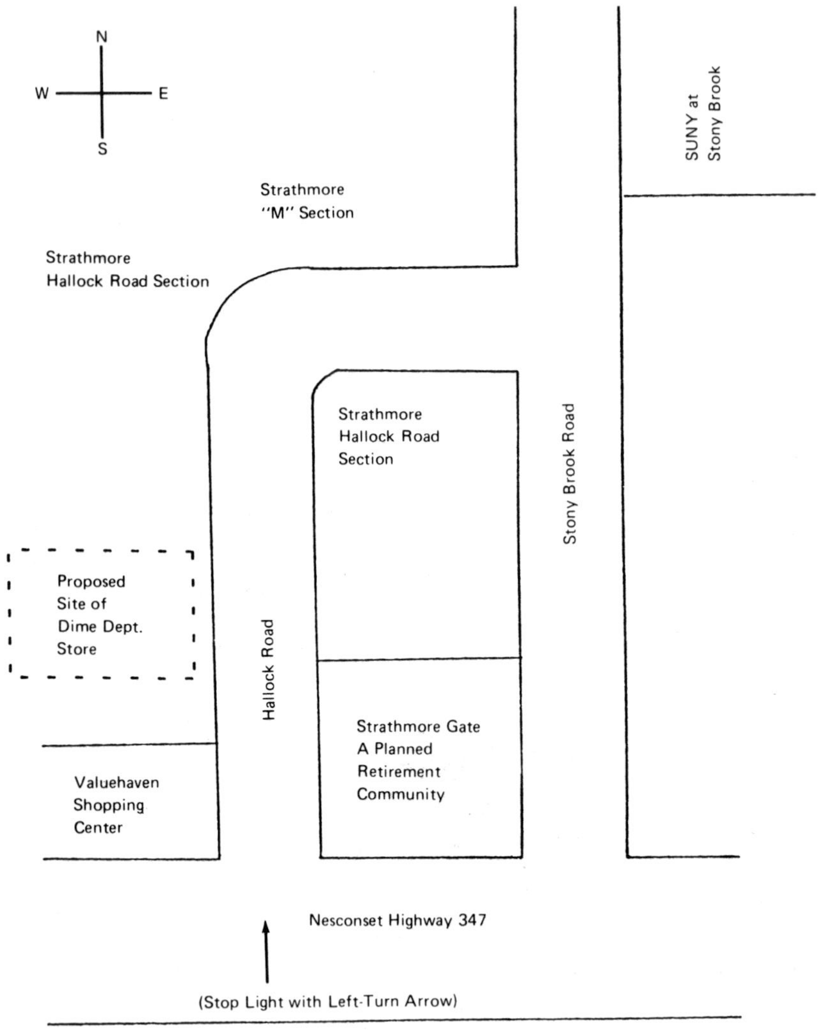

Figure 5: THE HALLOCK ROAD AREA

lights, and in turn the latter residents signed a Hallock Road petition requesting traffic controls for their street. This understanding, however, dissolved shortly thereafter. As one resident remarked:

> By the time of the meeting, however, the "M" section had become 'paranoid' about the possibility that the Town Board would totally capitulate to the Hallock Road request and that they would get the severance of the road. So on the night of the meeting they sent a secret telegram to the planning board asking that their names be removed from the Hallock Road petition.[6]

When the town board convened to discuss the local township planning board's hearing and to decide the issue of area impact, they took advantage of the disunity present in the Strathmore project, and they totally rejected all demands and petitions. In particular, they said that stop signs were definitely rejected as a control of traffic, and that a road severance was too radical an idea to be considered at all.

Three years later, the town board was to completely reverse its position. Furthermore, by that time, the bi-county planning board's department of traffic safety would recommend as the only rational proposal a severance of the road, returning it to a residential character. In order to understand this ironic reversal we must examine the step-by-step unfolding of the drama as the temporary reality of 1969 underwent a continual process of renegotiation.

The new residents of the Strathmore development found themselves in a constant state of ecological readjustment to their environment from the very day they moved in. The apparent "taken for granted"-reality changed as Hallock Road became widened and paved. Through traffic now appeared in increasing density as the Valuehaven shopping center on its southern end and the SUNY parking lot on the northern end were completed. With each stage in the development of the area new "unanticipated consequences" occurred as disturbances in the every-day pace of community life. Mothers, who once allowed their children out to play with impunity, suddenly became fearful of their welfare because of the increase in speeders on the street. Weekends, which used to be spent in the relative quiet of suburban surroundings, were marred by the high density use of the road to and from the shopping center and the SUNY lot. Because an integrated plan for the area was nonexistent, none of the residents were capable of predicting these results before buying their homes. The changes severely altered the conditions under which they had made the choice to belong to the community and affected their future prospects for remaining.

The many environmental disturbances, however, had differential effects on the residents depending upon their physical location in the project. Some areas of the development had more stable surroundings than others. Not having been called upon as an entire community to deal with the projected problems of further growth of the area, the residents responded to each unanticipated consequence with only the narrow view of their particular interests. They reacted to what affected them directly and not necessarily to what posed a hazard to other streets in the community. Thus the Hallock Road and the "M" section people, while virtually living in the same neighborhood, had qualitatively different responses to the disturbance of increased traffic. The frame of reference that influenced each individual in shaping his response was a very local one.

Although the possibility of collective action existed in this suburban community, several obstacles had to be overcome, including the privatized view of the residents and the lack of any ongoing citizen association when the families moved in. These two aspects are somewhat related and follow from the nature of such housing. New suburban communities acquire homeowners almost simultaneously with the project's completion. The builder is involved in a private transaction with each family and terminates his commitment to the area once the homes are sold, and the streets are turned over to the township. A developmentwide community association must be structured by these new residents themselves, but this implies that neighborhood interests have to be articulated on a communal and not privatized scale. The interaction process by which such a voluntary association arises takes a relatively long time. In the early stages, residents have little connection with each other beyond the block on which they live. Face-to-face social interaction is localized, and the relations between neighborhoods within the community are limited and sometimes divisive. If collective action does occur, it tends to remain issue-specific and be associated with an ad hoc group. The Hallock Road residents remained in a very parochial organization and their concern with traffic hazards were not shared by other neighborhoods, who tended to view the Hallock Road citizen activities as a threat to their own peace and quiet.

The businessmen and political leaders took advantage of the social fragmentation in the community and ignored citizen concerns over the effects of the commercial project. The renegotiation of the costs residents were called upon to bear as a consequence of these decisions however, progressed beyond this initial stage and did not end with the citizen defeat at the planning board meeting.

The first interchange between public officials and residents of the area occurred on July 16, 1968 when a letter was sent to the police department requesting a speed limit sign on Hallock Road (at the time there were still *no* posted speed limits). The police responded on November 13, 1968 with a survey of the area. It noted that the developer had not at that time dedicated the roads to Privatown, hence no action could be taken. Thus, while Valuehaven was going up, Hallock Road was still not an official township street despite the formal departure of the developer after all the houses were sold. (See map of the area in Figure 5.)

Over the next year, a number of residents acting separately requested the police department for various traffic controls on their section of the road. In April of that year, survey results were made public, and speed limit signs were suggested as the only traffic controls:

> *Traffic Volumes.* Light traffic volumes, moderately influenced to some degree by commuter traffic.
>
> *Approach Speeds.* Hallock Road is within a 50 MPH state wide speed zone in the area investigated, safe approach is 40 MPH.
>
> *Conclusions.* An evaluation and analysis of all the factors specified in Section 207.2 of the New York State Manual of Uniform Traffic Control Devices indicates that stop signs at the above location would not be warranted.
>
> *Recommendations*
>
> 1. The Town of [Privatown] and the Suffolk County Department of Public Works petition the New York State Department of Transportation for an evaluation of the speed limit on Hallock Road.
> 2. The Town of [Privatown] Highway Department erect "30 MPH Area Speed Zone" signs and "End 30 MPH" signs on these roadways intersecting Hallock Road between Nesconset Highway at its southern end and Stony Brook Road at its northern end pursuant to instructions contained on the attached TSB-3 forms.[7]

Two interesting features of this early interchange are worthy of mention. First, the apparent reality of Hallock Road was that it was unpaved dirt and in that sense rural when individuals moved in. Yet despite its appearance, the road was *meant* by the state highway plan to be a through street within a 50 mph state speed-zone area. Appearances can be deceiving, especially when developers are not required to notify their customers of such government schemes. Second, the referrent for the police department's entire analysis of the survey consisted only of the New York State Manual of Uniform Traffic Control Devices.

Throughout 1969 and 1970, a number of requests were made by individual local residents for traffic controls. Each survey by the police department resulted in a restatement of the same conclusions listed above in the first report.

In 1970, another change occurred which drastically increased the problem in the area. The Valuehaven shopping center, now completed, received a left turn arrow at the intersection of the highway and Hallock Road, enabling cars to turn onto Hallock without interference from other cars in the opposite lane. This left turn arrow was granted as a matter of course, following a formal request, in order to prevent a bottleneck.

The solution had a crucial unanticipated consequence. At the north end of Hallock Road lay the SUNY parking lot. Riders—thousands of them—destined for the university, while previously indifferent to turning down any one of a number of streets off Nesconset Highway to get to the parking lot, now found it extremely convenient to make their turn on Hallock Road, and cars sped right on through once the turn was made.

A number of accidents subsequently occurred, most involving pet fatalities. The following letter was sent to the Suffolk County Legislator Walter Napier by a resident:

> Dear Sir;
>
> Something has to be done about the traffic on Hallock Road in Stony Brook. This is a residential area with small children, but it is becoming a thruway and a shortcut. Since the installation of traffic signals on the corner of Hallock Road and Nesconset highway, more and more people are using this road to cut through to Stony Brook Road and the parking field of the University, using speeds in excess of the 30 MPH limit posted all along this street. Any steps you can take to alleviate this hazardous condition will be greatly appreciated.[8]

Napier, a Republican was chosen since his district included the Strathmore development and, while the Republican to Democrat voter registration was four to one in the area, he obtained a poor showing among those residents in his last election. It has been rumored that he was quite receptive to doing some favor for the voters in order to regain their support.

Napier sent a letter to the town highway department requesting them once again to investigate the area, adding the following paragraph:

> However, in as much as you are a resident of the area and also the Town Highway Superintendent, you should be keenly aware of any and all situations having to do with town roads. I did not feel that

> it was actually necessary to send you a copy of Mr. Smith's letter; with the construction of the State University parking lot, plus the through road from Nicolls to Stony Brook road, conditions which have aggravated an already bad situation, I assumed it would have been apparent to you.[9]

Surely, Napier believes that someone must be responsible for coordinating residential with economic and social location. He is, of course, wrong. Had he pursued this line of inquiry with a search for a responsible planning authority through the labyrinth of bureaucratic organizations, the effort would have been fruitless. Once the town board decisions were made, coordination of growth was not planned. The process of adjustment which the legislator was now caught up in is an ex post facto renegotiation between all affected parties resulting from events which appear to "just happen."

On October 22, 1971, the police department reinstituted traffic surveys along this route. The surveys conducted by the same people and in the same manner as those of 1968-1969, continued on through February of 1972. Due to the redefinition of the existing situation to one of greater political importance with the presence and interest of a county legislator, and the increased traffic to the SUNY lot, the police traffic analysts came up with a slightly different recommendation. While still insisting that the traffic engineering principles found in Section 207.2 of the traffic control manual indicated that stop signs were unwarranted for the area, they did concede that an alternative existed:

> It is recommended that the Town of [Privatown] Highway Department reduce the width of the moving lanes by painting a simulated Mall approximately four feet to six feet wide down its center. This mall would then allow two eight feet parking lanes and two eleven foot to twelve foot moving lanes.[10]

In the earlier report, the traffic referrent was used as a legitimation for turning down the individual citizen's request. It appeared that uniform government principles and not discretion were used to make that decision. At the later date, however, while the same principles were in operation, an alternative had been found. The mall would effectively reduce the wide (55 foot wide) through street to a narrow residential one. The appearance of such an alternative did not occur until the renegotiation process had reached the point where both the presence of a county legislator and the actual increase in traffic made the residents' demands more legitimate.

Until January, 1972, the residents in the area including the fifty families living on Hallock Road responded individually to the situation. During that month, however, a fatal dog accident resulted in a number of neighbors getting together to discuss their collective interests. Since the letter to Napier nothing apparent seemed to be being done to remedy the traffic hazard, and they were upset. A local resident met with Mr. Smith, a neighbor who was a commentator on a radio network and was looked upon as the informal leader of the area. The two men decided to contact the other residents and hold a meeting about the problem. Most of the fifty families living on Hallock Road agreed to participate.

From the very beginning, the twenty or so people who showed up were divided between the more "radical" elements, who wanted demonstrations in the street that would close off the road to traffic and sever it as they desired, and the more "conservative" elements, who merely wanted to make a more organized effort to contact political leaders.

What followed at the first two meetings of the community group was an effort to define the nature of their association and form a common understanding of its purpose. Those not present in 1968 were made aware of the history of the situation by the older residents and were instructed in the role the "M" section people played. By the second meeting the group had reached agreement on issues. First, they wanted to avoid a divide-and-conquer situation and did not wish to alienate the "M" section. Hence, they decided not to ask for a severance of the road. Second, by this time the political naivete of the residents had worn off, and they knew that in order to get anything done they would have to deal directly with the political leadership of the township and any "higher-ups" in the party that would be receptive to their cause. As the leader indicated in an interview:

> At the second meeting we decided to form an organization that would be able to contact political leaders and exert *formal* pressure. We opened the floor to motions about a name. We rejected the idea of calling ourselves a Civic Association because we felt that our complaints were more serious and urgent. Finally, we decided on the name HARP, the Hallock Road Association of Responsible Parents.[11]

The people felt that although they were only a minority of the residents in the area, by creating an ad hoc formal organization, they could exploit politically the name of their group by getting petitions signed for their proposals, which would then be backed by a larger constituency.

They also felt that local politicians would be much more receptive to a formal community organization. Failing to obtain any response to an already unplanned, seemingly accidental situation from the local town board, which purports to coordinate such affairs, they were left with no alternative but to become politicized.

In trying to consider what to ask for, the group attempted to act in a reasonable way to alleviate the situation. Far from there being a potentially infinite variety of choices, which they initially believed to be the case in 1968, they were compelled to agitate for the alternative that held out the greatest chance of being acceptable, traffic controls using stop signs.

> Given the existing constraints, I asked the people, "what is the most *reasonable* logical thing to ask for that would ease the traffic situation?" After deliberation we all agreed that stop signs seemed the only thing that we had any possibility of getting which would work.[12]

At the third meeting of HARP, the entire area had been informed of its existence. The members decided to hold a protest march primarily for the purpose of gathering publicity to back up the demands sent to the political leadership. At this time, a number of the residents objected to this militant action. A suggestion was made by the conservative part of the group. They felt that a letter sent directly to the local political leader, Carl Middleton, would be enough to bring some results. Hence, they split from HARP and retired from the dispute to write their letter.

> Numerous attempts to communicate to our local government the fears of our citizens for the safety of their children, themselves, and their property, have met with lack of response and as it appears to us, lack of concern. Members of the community are meeting with frustration and are beginning to believe that militancy is the only viable solution. However, we believe that local government desires to be effective and that working with it is in the best interests of all.
>
> As you have the means, and we may add, the obligation, to bring just relief to your constituents, we expect that you will participate with us in the solution of this problem. It is of the utmost urgency that you or your designate meet with members of our community at the earliest possible time to discuss concrete plans for a solution to this matter.[13]

To date, the senders of this letter have not received any reply. The failure of the township political leader to answer the conservative group encouraged HARP to continue with its own more militant pressure tactics.

In January, HARP submitted a formal request for stop signs, and in the days directly to follow, they demonstrated peacefully on the street blocking it for a long enough time for the incident to be reported to the police as a minor disturbance. They sent a publicity release to the local newspapers and received favorable press coverage in the *Three-Village Herald,* the local community newspaper. Before the demonstrations, they all agreed that a *neighborly* attitude should prevail so as not to upset other residents, in particular those of the "M" section. No formal response to the HARP request was made by the authorities.

The February series of traffic surveys were conducted at this time. The Suffolk County Legislator Napier sent another letter to the police department where he stated:

> It has been alleged that when members of the Suffolk County Police Department had been questioned, on the part of the residents, with regard to this problem, the reply was, 'The only way you will get any action is after someone gets killed.' Certainly, this, in my opinion, and I am sure you will agree, is not the proper response or method in dealing with the above matter.[14]

In the course of canvassing political leaders, Mr. Smith contacted New York State Senator Bedoya, who is the liaison between the local Republican party and the New York State Republican leadership. According to Mr. Smith, Bedoya was quite unreceptive to the HARP position until Mr. Smith mentioned that he was a newscaster for WNRX Radio. At this point, Bedoya seemed to change his attitude and said that he would see what he could do.

In February, Bedoya wrote a letter to Gary Bibby, the supervisor of the town board.

> Mr. Smith is a news commentator for WNRX. For one half hour he told me about the problems of Hallock Road being used as a speed zone to get from Nesconset Highway to Stony Brook Road, that a couple of kids have been injured, and that there have been several accidents involving climbing on lawns and etc.
>
> May I have your comments as to what we can do to help this group?[15]

This letter was also sent to the highway patrol, and Mr. Smith has reason to believe that it was widely circulated among the various political leaders of the town, who immediately thereafter displayed an awareness of his job. However, Bedoya was informed by Bibby that as supervisor he could do nothing since the traffic department had not recommended any controls as a result of their surveys.

Mr. Smith, now president of HARP, answered Senator Bedoya referring to township supervisor Bibby's letter and closed with this paragraph:

> Senator Bedoya, I am quite aware of the political realities of granting our wishes. The decision does not rest with supervisor Bibby, with Town Board or indeed with the police department, but with the Republican leadership in [Privatown]. Carl Middleton, the GOP town chairman, has been totally unresponsive to our requests. In fact, Senator, he has ignored us completely. Such an attitude is intolerable.[16]

The letter emphasized the fact that HARP was aware of the political realities of the township and was calling for negotiation with the seat of power, and not the legitimating safety valve forums operating there.

Shortly after, despite the new recommendations of the traffic survey calling for the construction of a mall, Bedoya sent the following to HARP:

> I have taken the liberty of speaking to Councilman Robert Clark and Councilwoman Ellen Buchanan.
>
> Councilman Clark has agreed to sponsor the ordinance setting up stop signs on Hallock Road, and I am sure that Councilwoman Buchanan will go along. It looks fairly certain that Councilman Clark may get the necessary majority to place stop signs on Hallock Road.
>
> I must point out, however, that I feel you are in error in your assumption that the Republican organization has anything to do with Mr. Bibby's decision to go along with the recommendations of the Suffolk County Police Department.
>
> I know Supervisor Bibby for many years and he is a very honest, diligent and hard working public official who calls the shots as he sees them. In this case, he prefers to go along with the so-called expert, who is the Suffolk County Police Department, and truthfully, there is a lot of merit in his position.
>
> Councilman Clark feels that since this road is a connecting road to two main highways that it makes little difference whether traffic is

> slowed up or not, and in the interest of safety would prefer to slow the traffic pattern in this area.
>
> Let me again assure you that there are no "political realities" in connection with your request, but actually one that must be weighed between traffic flow and reasonable safety.
>
> The attitude of both Supervisor Bibby and Councilman Clark are correct. It is a matter of choice.[17]

In order to understand Bedoya's remarks, two factors must be explained. First, since the Republicans held five out of the six council seats at that time, town board decisions were often made in private session with the township boss, Middleton. Bibby although a Republican, was somewhat independent of this process as he was the town supervisor and did not usually attend such meetings. For the councilmen, however, the meetings are important because they may decide whether to support the proposal, as well as how each individual member shall vote. Thus, an individual councilman may take the opportunity to champion the cause of any group without jeopardizing the ultimate outcome of the board's decision. This situation is taken advantage of especially by incumbents up for reelection. In this particular case both councilman Clark and Buchanan were up for reelection. Furthermore, it was well known by HARP that Clark had been trying to build up a constituency among the more liberal Republicans in the town. Hence his support for the proposal played to his advantage among the Hallock Road residents.

Second, the attitude of Bibby was a result of his own background. For many years before becoming supervisor, he was Township Highway Superintendent. In the years at this job he displayed a desire to make decisions totally by the book, the engineering rationality embodied in traffic manuals.[18]

During the month of March, HARP received a feeler which indicated that they would receive two sets of stop signs, but at a subsequent meeting the residents decided to turn the offer down. They were encouraged by the overall response from the board, and they felt that two stop signs would be ineffective in slowing traffic along the mile-long strip. In addition, they were still not convinced that the board would rule in their favor, especially after contacting Bibby on a number of occasions. They agreed, instead, to press for a maximum demand of seven sets of stop signs—one set to cover each intersection along Hallock Road. Councilman Clark was invited to the area, he took a "walking tour," and then agreed to sponsor the "seven-stop-sign" amendment.

At this time, HARP made a highly beneficial connection with the elderly people who live at the intersection of Nesconset Highway and Hallock Road in a planned retirement community called Strathmore Gate. It was located opposite from the Valuehaven shopping center, and the retirees complained that they couldn't cross Hallock Road safely in order to get there. Many of them were geriatric cases with aluminum walkers and canes. It was decided by HARP that they would request the seventh set of stop signs to be placed on the northern end of the entrance to their community. In return the elderly people agreed to sign petitions backing up HARP's resolution. The formal organization of HARP was thus used by this coalition of various citizens.

A few days later, Councilman Clark introduced the stop-sign resolution calling for the seven sets, and a date was set for a public hearing. Mr. Smith asked to be recognized at the meeting and requested that a separate hall be hired for the evening since HARP expected to have over 200 people in attendance. Later, Mr. Smith admitted to me in an interview that this was a bluff. Bibby suggested that it wasn't necessary for everyone to show and that a signed petition would suffice. Mr. Smith then agreed to consider that as an alternative.[19]

During the previous week, the members of HARP went around with their petitions to the other residents of the development. They spoke to neighbors previously never contacted because they lived in sections not affected by the traffic. These residents were predisposed in favor of the status quo, due to the convenience of having a through street to save them commutation time. Nevertheless, the fifty or so members of HARP got a number of other signatures.

As a result of diligent canvassing HARP was able to appear at the hearing with a petition of 250 signatures from Strathmore with an additional 75 from the private retirement community (PRC) of Strathmore Gate. As far as the town board was concerned there was no reason to doubt that most of these people were regular members of HARP, and the group of not more than 50 people were able to capitalize successfully on their formal structure and exploit the power of their position.

In addition, Mr. Smith had made it known that he was upset by the "bossism" he found in Privatown, and that he was planning to do a radio series on the situation. He too exploited his position and approached Mr. Clark about the possibility of him appearing on one of his shows. Clark, in fact, consented to be interviewed.

At the public hearing the residents of HARP backed into their corner demanded seven sets of stop signs. The police traffic department protested and argued against such irrational controls. Two weeks after the public hearing, the town board voted to place the full seven sets of stop signs on the street. Two "no" votes were cast, one by Bibby and one by Councilman Hymer, the senior member of the board, but the resolution was passed five to two.

About a month later a total of *eight,* not *seven,* sets of stop signs were installed overnight along Hallock Road!

I asked Mr. Smith if there was any intention of disbanding HARP. He said in June there was such a suggestion brought up at a meeting. But the people decided that despite discontinuing the weekly meetings they were sure that if other contingencies would come up, it would be nice to have some kind of formal organization to handle them. The ad hoc group now appeared to be taking on more permanent status, however, HARP was much removed from the general backing required for a community-wide civic association.

He then mentioned that, in fact, some of the people believed that, perhaps, the situation has not played itself out yet. It appears that despite their precautions there is much ill feeling in the "M" section about the stop signs. There is a rumor that people there are trying to organize themselves for a petition drive to *get the signs removed!* They claim that not using Hallock Road makes it longer for them to go to work or commute through the area. Thus, the renegotiation process continues.

The final irony of the situation is supplied by the bi-county traffic safety department's $5,000 survey requested by Napier and made available in May. After an extensive analysis of the situation, it recommended that since Hallock Road is a through street anomaly in an otherwise residential area, the road should be severed and returned in two parts to a residential environment.[20]

Citizen Participation and Party Vulnerability

The present landscape of Privatown is taking on a shape which is difficult to explain. The individual finds that he has little control over his community environment. This occurs even when residents attempt to make meaningful choices during political elections. Many events that occur in his local area, such as the development of a large shopping center, seem to "just happen" without connection to the rational use of land or benefit from local government coordination. Explanation for this *unanticipated* aspect of everyday life is compounded by the paradoxical existence of planning boards, township public hearings, departments of traffic, and professional planners, all purporting to regulate the exigencies of social growth. Despite their presence the individual remains without effective public agencies for controlling or predicting the process.

Given the unresponsiveness of political leaders to local desires, there is a growing tendency toward legal action. Citizens can go to court to reverse a government decision or sue for damages. Such action requires financial resources and wide citizen interest, consequently, fighting "city hall" after decisions are made is not a frequent occurrence. In the more developed region of Nassau County, however, legal challenges to township decisions have increased in recent years, at times running as high as 30 percent of all re-zoning decisions, such as in the township of Port Washington. In Privatown, during the four year period we have studied, local citizen groups initiated court action only twice. Both of these attempts were successful and were sponsored by relatively wealthy sections of the township.[21]

In most instances residents still respond to the environmental effects of business and government activities and not to the land-use decisions themselves, because they can do little about these once they have been made. This was the case with Hallock Road. In this instance the initial cause of community problems, the down-zoning and subsequent development of the shopping center, could not be challenged practically by the residents, because it was already being constructed. Instead, the people in the area attempted to renegotiate the social costs of such growth by responding to its neighborhood impact. This response is limited because the public is relatively unorganized. There is a lack of any authoritative definition or shared meaning of the concept "community." Residents express loyalty to their immediate neighborhood, but hold only a vague notion of the patterns of growth, social needs, and characteristics of the larger suburban township within which they live. This "limited liability" is reflected in the narrow interests expressed by groups attending town board hearings. Civic associations and school district representatives appear at meetings whenever their own immediate area is affected. After assessing the relative impact of a proposal on their communities, they send spokesmen and/or petitions to the town board so that their position can be publicly recorded. In cases where local civic associations do not exist in an area ad hoc groups are generally formed. These temporary associations also arise in response to specific proposals.

In Privatown there is no mechanism which allows residents to articulate their own specific needs or combine with other residents in formulating *aggregate* demands in order to advance the public interest. Citizens usually assume that through the operation of home rule, town board meetings and public hearings, such an integrating function is already being carried out by the local political leadership. As we have seen, the importance of broad, coordinating political leadership has all but been ignored in favor of a more limited, privitized version of the public's needs.

However, there *are* a number of voluntary groups which have attempted to work in the public interest. There are environmental groups, such as the Suffolk Defenders of the Environment, "good government" groups, such as the League of Women Voters, and civil rights groups, such as the Privatown Housing Coalition. These local organizations monitor the decision-making activity of local government from the perspective of county-wide needs and attempt to represent broader based interests than ad hoc or civic associations. Whenever a relevant decision is being deliberated by the town board, they send a representative to the public hearings to express their positions. The League of Women Voters has a representative at every Privatown town board meeting who attempts to insure that the councilmen follow legal procedures, who argues for greater public awareness, and who keeps an accurate record of the board's deliberations. These public interest and "good government" groups, however, have a limited impact on political control and have not been able to generate wide public support needed for good planning. Their last refuge, like civic associations, is to challenge political decisions in the courts. Recently, such groups have also become more politicized as their list of countywide concerns has grown.

Whenever resident sentiments become more organized into formal groups, they tend, for the most part, to operate with very narrow interests and a limited time horizon. Local citizens are pressured to consolidate their positions within a formal, countervailing power group and play the bargaining game of interest-group liberalism. The Hallock Road residents were unsuccessful in calling attention to their needs simply by separately contacting their local political leaders or attending public hearings. Response was made only when they had marshalled the threat of organized political power. At this stage, however, they had to enter a renegotiation "game" and bargain with other organized interests. This outcome of interest-group liberalism is a formless one and replaces broad social goals with bargained outcomes. As Theodore Lowi has argued, a government which relies on pluralist negotiations to implement decisions instead of the planned selection of social goals cannot carry out strong, positive programs that directly satisfy social needs.[22]

The outcome of the Hallock Road renegotiation process, which involved these elements, was one which nobody wanted. The irrational use of stop signs became the only reasonable choice once the political interest-group game had begun. Such a process substitutes a formless procedure of pluralist bargaining for a planning approach to resident and business needs that would determine social goals through the public articulation of broad social interests. In the absence of such a mechanism, the outcome of the present pattern of political response may very well be the organized politicization of every single interest in the society as the unanticipated environmental effects of growth continue to proliferate.[23]

NOTES

1. The covenant is, then, a private agreement between the developer and the public.

2. Minutes of Planning Board Meetings, 1967.

3. Ibid, March 7, 1962, quoted in Howard Scarrow, "Political Control of Zoning Decisions," in Dieter K. Zschock (ed.), *Brookhaven in Transition: Studies in Town Planning Issues,* Stony Brook, N.Y.: SUNY at Stony Brook, 1968. Chapter 4 is based upon the work of Scarrow, the first to study the political use of zoning to support the local party.

4. This and all subsequent quotes of local residents are taken from case study interviews and letters from HARP—the ad hoc association of Hallock Road residents.

5. Ibid.

6. Ibid.

7. Suffolk County Traffic Department Survey, 4/23/69.

8. Case Study letters, op. cit.

9. Ibid.

10. Traffic Survey, 10/22/71.

11. Case study, op. cit.

12. Ibid.

13. Ibid.

14. Ibid.

15. Ibid.

16. Ibid.

17. Ibid.

18. The New York State police traffic manual mentioned as the referrent in the previous surveys of the area heartily rejects the use of stop signs to control traffic. They are suggested to be used only in the case of two equal sized roads that transect each other. Whenever unequal flows intersect, other measures are suggested to be used. In addition, signs are *never* to be used to slow down traffic flow.

19. In support of the public hearing county legislator Napier asked another bureaucracy, the traffic and safety department of the bi-county planning board, to make an exhaustive survey of the situation and to present its own recommendations. While

the results of this survey, which cost an estimated $5,000, were not needed by HARP because they appeared after the petition was granted, it added an ironic element to the case study.

20. This was the same position advocated by unorganized residents in 1968.

21. The first was the case of "DePoppas vs. Barraud" in which the awarding of a re-zoning from residential to business for "7-11" Store in July 1970 was declared void by State Supreme Court Justice Henry Tasker. The judge ruled that it was a clear case of "spot zoning." His decision was appealed by the board, but was upheld by a three to two decision of the appellate court. The case is now before the court of appeals in Albany. The concept of "spot-zoning" used by Tasker in his opinion refers to the flagrant re-zoning of a single parcel of land to a category clearly incompatible with that of a surrounding area. Thus, the "7-11" Store, which has continued to operate pending appeals, is in the middle of a residential development. The second case was a ruling against a town board re-zoning which gave the petitioner permission to build an office building in an otherwise residential area. Tasker ruled against the favorable re-zoning obtained by a relative of Louis T. Juliet, the chairman of the Privatown Zoning Board of Appeals.

22. For pluralism to work in political negotiations there seems to be a minimum requirement that interest groups be relatively organized and broadly based. Certainly, in Privatown the parochial, ad hoc nature of most residential associations greatly limits the ability of pluralist bargaining to be effective, even if we disagree with Lowi and claim it to be a viable decision-making mechanism.

There is some indication, however, that pluralist decision-making is not working effectively in more mature, developed regions of the country. Cf., Theodore Lowi, *The End of Liberalism,* New York: W. W. Norton, 1969; Robert Paul Wolff, *The Poverty of Liberalism,* Boston: Beacon, 1969; Todd Gitlin, "Local Pluralism as Theory and Idology," in Hans Peter Dreitzl, *Recent Sociology,* No. 1, New York: Macmillan, 1969. In contrast there are many central city studies which still claim it to be effective: cf., Robert Dahl, *Who Governs?,* New Haven, Conn.: Yale Univ. Press, 1961; Edward C. Banfield, *Political Influence,* New York: Free Press, 1961. Much of this argument, moreover, is part of the larger question of political control and is expressed in the plethora of articles, books and case studies on the community power debate. Cf., for example, Hawley and Wirt (eds.), *The Search for Community Power,* Englewood Cliffs, N.J.: Prentice-Hall, 1974. These studies do not address the needs of the submetropolitan polity.

23. Without wishing to add to the community power debate, the Hallock Road study indicates the need for new decision-making mechanisms and is not a defense of either elitist or pluralist notions of power. The fact that interests are so fragmented in an otherwise relatively homogeneous area supports this view. Few ovservers have volunterred reconstruction schemes to help aggregate effectively local community sentiments. For a discussion of this issue, cf., Gerald D. Suttles, "The Search for Community Design," in *Metropolitan America: Papers on The State of Knowledge,* New York: Halsted Press (A Sage Publications book), 1975.

Suttles emphasizes the need for reconstruction of social structure and not the building of new physical ones as the necessary approach for planning efforts. After World War II, Karl Mannheim was also extremely concerned about viable decision-making design and reconstruction as a consequence of the totalitarian political forms taken by European countries. Mannheim called the desired public aggregating mechanism which would democratically make decisions the "organic articulation" (Gliederung) appartus. Cf., Karl Mannheim, *Man and Society in an Age of Reconstruction,* New York: Harcourt Brace & World, 1940; and *Freedom, Power and Democratic Planning,* London: Routledge and Kegan Paul, 1951. The former contains an extensive, valuable bibliography on the study of democratic social planning techniques. For an American view of Mannheim's approach, cf., John Friedmann, *Retracking America,* New York: Doubleday Anchor, 1971. For a summary of the numerous economic problems involved in making democratic choices with public goals, cf., Sidney Hook (ed.), *Kenneth Arrow's Welfare Economics: A Symposium,* New York: New York Univ. Press, 1969.

Chapter 8

PRIVATOWN AND THE POLITICAL RECONSTRUCTION OF SUBMETROPOLITAN AMERICA

The Emergence of Submetropolitan Regions

The patterns of urban growth in the United States are often based upon migratory trends. The period of city growth and industrial development in the 19th century was characterized by the great influx of immigrants. Today's problems, however, are caused less by the rapid growth of central city population and more by the massive spread of people, housing, and economic and social activities in progressively maturing urban regions of increasing scale. The centrifugal patterns of development have been at work largely since World War II, but only recently have we come to realize that there is a new urban form of settlement which presently characterizes American life. These are the somewhat autonomous metropolitan regions that are associated with at least one, but often several, central cities, such as the New York or the Minneapolis-St. Paul metropolitan regions, which extend for hundreds of miles and comprise a varying landscape of cities, towns, sprawling suburban developments, farms, and industry.

The population spread outside the central cities includes a majority of nonfarm residents and will continue to do so in the future. We have called attention to these separate suburban developments as aggregating to comprise maturing subregions within the expanding metropolitan mosaic. These suburban, submetropolitan areas are increasing in complexity, interdependence, and declining in subordination to the central city. At the same time, however, the entire region, including the city, has become more dependent upon national patterns of bureaucratic and corporate social control, accompanying a general increase in organizational scale. These nationwide networks of hierarchical social organization and development are the consequence of the industrial expansion of both business and government along with their associated communications and transportation technologies.

Articulating the dimensions and patterns of the new urban subregions is only the first step in meeting the challenge of their needs. At the national level regional suburban growth has been encouraged by federal legislation supporting the construction of housing and highways. While the sheer volume of post World War II home construction is impressive, the inequities of such growth are becoming more apparent. Presently, the suburban, submetropolitan areas are encountering local municipal problems and patterns of internal differentiation which progressively resemble those of the central cities. Privatown, for example, has fiscal difficulties in financing schools and in supplying much needed social services, such as adequate sewage treatment, while it burdens its residents with an increasing property tax rate. It has also accumulated the inequities of racial and income segregation, inadequate housing and blight, spread city sprawl patterns of growth, the disappearance of open space, and environmental problems of pollution. Its residents are finding it increasingly hard to meet adequately the subregional needs and the problems created by the centrifugal land-use patterns of development.

The limited success of massive subregional settlement is not so much due to rapid growth as to the absence of political control and regional coordination. Suburban areas suffer from lack of internal coherence and consequently of the ability to regulate and shape development. This is so despite a large investment in planning and coordinating agencies at all government levels. In addition, the continuation of the revered principle of local home rule and political control over zoning does not seem to achieve the much needed regulation of land-use. Weak party control at the local level has combined with a failure of federal initiative along with limited voluntaristic efforts of suburban residents to provide the expanding submetropolitan regions with the mechanisms of coordination needed to shape the landscape to meet social needs. In Privatown, in particular, we have shown that the weaknesses of party organization have subverted planning efforts even when they have been tried, so that suburban sprawl continues as if land-use regulations do not exist.

No matter what future investment this society decides to make in planning programs and regulations, these may just provide local political control with more power over resources and more opportunities for party and personal gain. Similarly, future federal schemes may just supply the growing bureaucratic infrastructure of planning and the professionals working for it at all levels with more power and resources to support an elitist and physical style of planning prevailing in the nation. At this stage, therefore, a larger government investment in social planning without needed social reconstruction may not necessarily be better, and, in fact, may prove to be far worse than no response at all. What is absent and critically needed is first, the recognition of the suburbanized area as a somewhat separable but interdependent part of the larger metropolitan region. This fact must be appreciated by a social science theory of urban life which needs to retool concepts away from city-suburb contrasts and towards large-scale interdependent metropolitan regions. Second, there is a need for new mechanisms and governmental forums that are appropriate to the suburban, submetropolitan region. Broad public debate would be appropriate in drawing the required outlines of the reconstructed community design and regional political infrastructure. This dialogue was already begun by the local control movements of the 1960s and by the growing national emphasis on populism. Such efforts must be broadened and followed by long-overdue social reforms. The public response in articulating and implementing the necessary decison-making apparatus to fit the emerging submetropolitan regions needs to satisfy two overall aspects. First, it must enable these massive, interdependent areas to make better use of existing planning tools in order to coordinate growth without a "beggar they neighbor" competition over scarce resources or further expansion of hierarchical bureaucratic levels. Second, it must enable local citizens to present demands to planners and officials in a way that would bring adequate response to needs, and not further disenchantment. Such reform, therefore, would avoid feeding the already growing political cynicism of residents and make politics work at the street level of suburban cul-de-sacs.

The Contributors to Suburban Sprawl

In any urbanized area there exists a large cast of actors with a wide array of interests and incentives motivating their social decisions. These are invariably linked to perspectives which concern aspects of everyday life associated with the eco-system of the particular form of human settlement, from tribal village, rural town, city, to metropolitan region. Privatown contains an interacting mixture of commercial businessmen, builders, bankers, developers, speculators, politicians, planners, homeowners, and tenants. Unlike the central city, submetropolitan regional concerns center around housing that is relatively new but presently maturing; highways and commercial strips, which supply the concrete grid for the automobile culture connecting individuals and everyday activities; and a low-density community life of privately oriented residents uninterested in the constant contact offered by the more racially and economically heterogenous central city.

Most of the residents of Privatown are single-family homeowners living in suburban "developments," not communities, located in unincorporated areas. While housing is presently expensive, these residents have, for the most part, purchased new homes at a relatively reasonable price and have watched housing values appreciate substantially over the years. On the whole, therefore, the middle class, whose concern over housing is limited to material aspects, has found great opportunities for personal gain in the region, despite its present changing character and the growing interest in the quality of community life.

Developers and builders of housing and suburban-style shopping malls have benefited also by large-scale submetropolitan growth. The flat Long Island geography is ideal for the mass construction of homes. This is especially the case because sewage disposal and water supply were not substantial burdens on builders at the time of greatest growth. Local regulations left these matters to be dealt with by the future residents, who presently are encountering increasing fiscal costs and higher property taxes as a result. Township regulations, however, regarding building codes and zoning ordinances have constrained developers in two important ways. On the one hand, they have made them dependent upon local politicians for the approval of building projects. On the other, they have raised the costs of housing, made it more conformist or uniform in design and have helped to limit individual home construction and encouraged building by large, mass-market companies. In general, home construction has proceeded successfully within these constraints to produce a remarkably large number of dwellings for the millions of residents migrating from New York City and the suburban areas more adjacent to it.

Financial and real estate speculators along with a select group of local politicians have also benefited by the past three decades of submetropolitan growth. Their interests are not those of the commercial businessman or builder who makes profit by producing or providing some services. Instead, the former group is motivated by the opportunities afforded by rapid growth and the multiple increase in the value of land when converted from farms to housing plots or strip-zoned highway construction. Motivated by large-scale profit-taking in the rapid turnover of land, speculators set the tenor of growth by manipulating the political control of future land-use, so that, presently, they alone seem in command of such long-range interests.

Professional planners and bureaucratic government agencies are motivated by the desire to implement the physicalist schemes of master plans. Privatown residents are blanketed by such activities beginning with their own local township planning board and the Privatown master plan, and extending to the expanding hierarchy of government planning agencies at all other levels. The most ambitious scheme is the Nassau-Suffolk bi-county master plan, which involves the entire region. These efforts, however, are limited by several factors. First, and principally, they are advisory in nature. Local politicians retain political control of zoning, and, therefore, township land-use. Second, professional planners ply their trade in elitist ways, and they must engage in public relations efforts to win popular approval for their schemes. Resident support, if forthcoming at all, is reluctantly given. This is so because planners assume both the presence of easily aggregated public demands and the intrinsic appeal of physicalist designs when displayed in multicolored maps. Neither of these beliefs seem warranted. Presently, planners on Long Island have moved to overcome these factors by working directly with political and business leaders. This strategy continues to provide residents with reasons to distrust them further.

POLITICAL LEADERS

On the one hand, the fragmentation through small, relatively new developments in Privatown prevents residents from engaging in more socially oriented activities of regional concern. On the other, the weaknesses of the local political organization and the polity of Privatown produces no political leaders with broad public interests or orchestrated party power. Consequently, the combined activities of builders, speculators, homeowners, politicians, and planners have melded over the years to produce a sprawling, multiproblematic landscape which lacks social comprehensiveness and political mechanisms that would allow for the effective and equitable shaping of submetropolitan growth.

At the most basic level, the local political organization of Privatown is weak and ineffective. With some notable exceptions, submetropolitan regional political life is a dead end. Its participants are almost invisible outside of their own party functions. Despite the residency of talented, energetic business and social leaders, Privatown political parties seem to attract individuals with narrow concerns and limited regional horizons. Even the Republican party machine of Carl Middleton, which has dominated the township since the 1960s, exercises weak control over its constituents, and continually fights reform battles with other members or parties.

The local polity of home owners remains fragmented in small developments and compartmentalized by automobile transportation and mass marketing suburban malls. It is relatively satisfied with minimal township governance, despite growing discontent as the costs of everyday life increase to meet the neglected longer-range inequities of regional development. Consequently, local residents provide little support to the local party organization.

Patronage is not in great demand by the majority of homeowners, so that party positions and favors are ineffective rewards for maintaining party control and promoting success at the polls. In addition, the progressive decline in active party membership with growing voter apathy or independence leaves political organizations with financial problems in supporting themselves. We have seen that such factors have made party politicians heavily dependent upon the real estate and construction industry for money. This occurs primarily because local political control of land-use and building codes, supported by the ideology of home-rule, are the basic favors that political figures can grant.

Privatown politics can be characterized best in terms of the activities of a select group of local party organization leaders, speculators, and developers who operate both to support the party and to realize personal gain from rapid regional growth. The majority of businessmen outside this group compete with each other for choice locations along commercially strip-zoned roads, and thus accepting with limited concern the sprawl patterns of development. At the same time, residents show their limited interest in party politics in the growing number of independently affiliated voters. The presence of periodic, almost cyclical, outbreaks of corruption scandals, has stimulated little voluntaristic effort at reorganizing local township government by either businessmen or residents. Presently, the Privatown government is controlled by a reform slate of Democrats swept into office by the Watergate backlash and the more recent evidence of ineffectual Republican party control. However, the current Democratic leaders must also face the weaknesses of limited public patronage, the increasing independence of voters, and the growing expense of elections, which forces them to rely upon the businessmen most interested in utilizing influence to exploit political control over land-use. Consequently, the prognosis for more enlightened local government is not promising. Long overdue political reforms at all levels of the metropolitan region are necessary to reconstruct suburban government, and they cannot be accomplished within Privatown alone.

PLANNERS

With a weak political regime dependent upon using control of construction and zoning as primary sources of financial support, planning for more "rational" land-use has little chance of success in Privatown. Mindless spread-city patterns seem to prevail, even with the existence of master plans and local building regulations. In order to transcend their ineffective advisory role, professional planners have attempted to play a political game of influence. On Long Island this activity is of a nonpartisan nature in the sense that the planners attempt to influence either political party. In their effort to influence the growth patterns of the island, they fall back upon their own traditions of physical planning. Their mandate from each of the county legislatures, however, allows them to engage in collecting demographic information and in formulating the regional master plan. Such proposals, however, are usually subverted because local home rule townships and their residents do not support the mandate, nor are they involved in the planning. Master-planners, therefore, use public relations efforts and political influence to gather local support for their land-use proposals in each separate township of the region, regardless of which party is in control. At the same time they seek aid from higher levels of government, state and federal, for broader powers of review and implementation.

To date, these professional, nonpartisan efforts have had limited success. The elitist approach of planners and their use of influence among the business community, such as in the Oak Forest project, attract little interest among the people because they are left out. Residents are suspicious of the planners' motives and reject their advertising technique efforts at public meetings, because they have not participated directly in the schemes placed before them. Even advocacy planners, working for socially active and religious reform groups, such as Burns of SCDC, become targets of suspicion as projects are proposed which increase the equity of segments of the community. The planners' use of influence upon either politicians or businessmen for project support creates public suspicion towards the planners, precisely because the men they associate with have power and are also distrusted by the local residents.

Despite the entrapment of professional planners in the elitist/populist dilemma, there is indication that support for some planning has grown. The growing problems of land-use, highway congestion, housing blight, environmental pollution, and business progress are turning numerous residents of the region towards planners for help, and away from traditional, privatized, narrow community concerns. Even in this more favorable situation, however, professional planners are still limited by their own physicalist heritage. All they can offer the minority of citizens oriented toward regional coordination are the standard proposals of the master plan derived from landscaping planning theory. Adequate environmental arrangements for submetropolitan residents will not emerge from the implementation of these controls alone (the physical fallacy). Furthermore, even the most rationally engineered project, such as a new highway, can still be appropriated by the land-use control of local governments for use in acquiring party or personal gain. This was the case with Nesconset Highway which began with limited-access status, but became a strip-zoned speculator's bonanza. We must recognize, then, that reform of professional planning approaches along with national land-use legislation must be an *integral* part of any program attempting the political reconstruction of the metropolitan region.

Given the weaknesses of local political control and of professional planning in Privatown, centrifugal patterns of growth have riddled the area with inequities. Over the past three decades, there has emerged a growing maldistribution of resources. First, communities are segregated by income and race. The value of housing has appreciated in the wealthier areas of the township, while the less affluent sections have become the sites for suburban slums and blight. The absence of balanced settlement patterns on a regional scale is apparent in the inequitable distribution of social services, such as medical care and education, which seem to favor the wealthier communities located along the north shore, to the disadvantage of citizens at the center of the island.

Second, the privatization of land as single-family back yards on a massive, sprawling scale has led to the disappearance of farms and open space, and little or no use of land previously farmed but now held fallow for future building. This has created several environmental problems and a growing concern over water quality, sewage disposal, and the preservation of natural areas. The speculators' devouring of farms and the pollution threat of spread city growth touched off a reaction among the rural residents of the region's east-end, who attempted, but failed, to separate themselves from Suffolk County. All residents, however, are presently aware that environmental inequities will be paid for by them in the form of health hazards and higher taxes. Recently, for example, Suffolk County has had to embark on an extensive sewer construction program caused by lack of planning foresight and limited building regulations, which only required primary treatment cesspools. This has greatly raised the property tax burden of residents, and the project's escalating costs have had to be met by the use of federal funds and county deficit spending. To date, the project is continously encountering fiscal difficulties, it remains uncompleted and foreshadows greater national government involvement and more debt before it is finished. Home-rule, privately-oriented residents seem destined to face future limitations on local governmental control as they become forced to appeal to federal agencies for fiscal help, thus duplicating the central city experience.

Third, spread-city patterns of growth have created dependence upon inefficient means of housing, transportation, and land-use. The actions of spot builders, who have filled in the margins of space between suburban developments, along with the select group of speculators and politicians utilizing political control of zoning to make money from land, have aggregated to subvert planning guidelines. The mindless, wasteful sprawl has also been supported by local residents who have obtained area-wide upzonings of land to lower-density use in order to control uncoordinated population growth and to preserve the value of their homes. In Privatown, the landscape appears chaotic and congested, despite low population density and much planning. Traffic problems even make it difficult for individuals to purchase late-night single items from stores, such as milk and cigarettes. Total dependency upon the family car compartmentalizes neighborhood relations and interferes with the street life (ball games, bike riding, strollers, etc.) which once contrasted with city street life. Commercial strip zoning and sprawl patterns of low-density residential land-use, which presently characterize the Privatown landscape will make it extremely difficult for any viable planning attempts to be made in the future, even with a growing interest in such alternatives among the residents. This is the final irony which is just now beginning to be recognized throughout the region.

CITIZENS

As the area has matured, social inequities have become more evident, and citizens have become more conscious of the entire region. Local residents and voluntaristic organizations, however, find that this awareness has been expressed mainly in increasing cynicism and antigovernment attitudes rather than in greater involvement, because present activities appear to be self-defeating. This attitude is also expressed toward the federal highway and housing programs, which have failed to initiate appropriate measures of growth coordination, despite the availability of public funds. For Privatown residents more federal social planning may just mean further encouragement of regional growth and more bureaucracy, neither of which they want or need. They will probably not support such measures. At the same time, the cynicism and apathy of the residents is in contrast to earlier conceptions of a suburban citizenry highly involved in its township government.

Home rule has implemented resident low-density zoning needs, but has been ineffectual in molding growth, even with the responsive use of local town board political powers in land-use and construction controls. On the one hand, local government seems limited and self-defeating in its attempts to deal with the inequities of rapid regional development. On the other, the same patterns of growth have provided the local political organization with its opportunities to support itself, and residents are made aware of this through periodic land-use corruption scandals. On the whole, therefore, submetropolitan residents are becoming increasingly "turned-off" to the future possibilities of regional life and government social planning.

Even "successful" attempts at citizen redress of unpopular town board land-use decisions, such as in the Hallock Road case, illustrate the limited results which are obtained by active involvement. They show that local citizens are *re-actors* to business and local government actions, and can participate very little in decisions which effect their entire area. Experience also demonstrates that everyday life in Privatown can be characterized as involving continual adjustments to the emerging patterns of uncoordinated growth. Most of the time, these are made in a private way by individuals who justify each new environmental hassle as "just being a part of modern life." The Hallock Road citizens combined to renegotiate the social costs they were called upon to bear by "unanticipated" development. This involved them in a pluralistic bargaining process with numerous groups of individuals, such as traffic engineers, local politicians, and residents from adjacent areas in the same development, subscribing to divergent points of view reflecting their particular interests. Although the outcome of the bargaining met their demands, it involved a solution which is difficult to defend, and which nobody really wanted. Control of development which could have been accomplished through adequate planning was, thus, left up to a political negotiation process, where several conflicting points of view clashed and were compromised by *irrational* results. A resident from outside the area, unaware of the stop-sign controversy and driving up Hallock Road, would be subjected to traffic controls bordering on the level of a "practical joke," which contributes further to the incomprehensible suburban landscape. The closely placed signs signify no planning.

Reconstructing a political infrastructure to bring about coordination of the submetropolitan region, therefore, would require mechanisms which would prevent ineffectual responses by politicians and planners, and would insure public involvement so that needs could be met in an integrated way. The present political structure is so narrowly based that it trivializes civic life, attracting neither able political leaders nor wide-spread public confidence. Undoubtedly, social planning faces special difficulties in a society which heavily favors the ideologies of free enterprise and local control. But the submetropolitan regions surrounding our central cities have long since defied the myth that the local community is a world unto itself. They are vast areas of contiguous residential usage, industrial and commercial development. No form of metropolitan government will be able to govern effectively such areas unless they acknowledge the somewhat separate subeconomies of each region, and the need for widespread public involvement.

The prospects for a tiered metropolitan government, which would give explicit form to the subregions of our metropolises, may seem dim, and one is tempted to join the residents of Privatown in their cynical view that government itself is ineffectual. These same residents, however, are slowly surrendering the remnants of local control to large-scale, non-elective authorities (port and transportation, sewage, water, health care, etc.), which further reduce the scope of the existing local governments and the attractions of a career in them. Someday, suburban residents may come to realize that local government has nothing left to preside over except the subversion of zoning restrictions which the local politicians themselves help pass. At that time, the residents may recognize that they must recapture political control of those vast regional bureaucracies in a new political form that resembles the submetropolitan region itself. The paths toward submetropolitan government, then, may be quite indirect, and the further weakening of existing local governments a necessary step.

Certainly, the inarticulate debate among suburban residents will itself be a barrier to social change. Most of the residents and opinion leaders seem to blame big government and big business for the shapelessness of suburbia, and see in the further growth of government an additional obstruction to self-determination rather than an opportunity to construct an instrument appropriate to the task and broadly representative of the diverse interests present in our submetropolitan regions. The physicalistic orientation of planners and their doctrinaire adherence to text book principles only confirm the view of the residents. Slogans about "overcentralization" and "too much government" obscure the genuine weaknesses of local government as well as the fact that many public bureaucracies have escaped popular control. The hope, then, is not for immediate political reform, but for a more articulate public debate, which revolves less around the question of whether there will be more or less government than around the question of the appropriate form of government for these submetropolitan regions.

BIBLIOGRAPHY

Aleshire, Robert (1970) "Planning and Citizen Participation: Costs: Benefits, and Approaches," *Urban Affairs Quarterly,* Vol. 5, No. 4, (June), pp. 369-393.

Altshuler, Alan (1965) *The City Planning Process,* Ithaca, N.Y.: Cornell Univ. Press.

Babcock, Richard and Fred Bosselman (1973) *Exclusionary Zoning,* New York: Praeger.

Banfield, Edward C. and James Q. Wilson (1963) *City Politics,* New York: Vintage.

Bellesh, J. and M. Hausknecht [eds.], (1959) *Urban Renewal: People, Politics and Planning,* Garden City, N.Y.: Doubleday Anchor.

Berger, Bennet (1960) *Working-Class Suburb,* Berkeley: Univ. of California Press.

Berry, Brian J. (1975) *Growth Centers in the American Urban System,* Chicago: Univ. of Chicago Press.

Bolan, R. (1971) "The Social Relations of the Planner," *AIP Journal,* Vol. XXXVIII, No. 6, (November), pp. 389-396.

Campbell, A. K. and J. Dollenmayer (1975) "Governance in a Metropolitan Society," in Amos Hawley, et al., *Metropolitan America in Contemporary Perspective,* New York: Halsted Press (A Sage Publications book).

Clawson, Marion (1971) *Suburban Land Conversion in the U.S.,* Baltimore: Johns Hopkins.

--- and Peter Hall (1973) *Planning and Urban Growth,* Baltimore: Johns Hopkins.

Dahl, Robert (1961) *Who Governs?* New Haven: Yale Univ. Press.

Davidoff, Paul and Neil Gold (1970) "Exclusionary Zoning," *Yale Review of Law and Social Action,* 1, (Winter).

--- (1965) "Advocacy and Pluralism in Planning," *AIP Journal,* Vol. XXXI, No. 4, (November), pp. 331-337.

Delafons, J. (1969) *Land-Use Controls in the United States,* Cambridge, Mass: MIT Press.

Diesing, Paul (1968) "Socio-economic Decisions," *Ethics,* Vol. IXIX, (October), pp. 1-18.

--- (1962) *Reason in Society,* Urbana: Univ. of Illinois Press.

Djilas, Milovan (1957) *The New Class: An Analysis of the Communist System,* New York: Praeger.

Dobriner, William [ed.] (1969) *The Suburban Community,* New York: Columbia Univ. Press.

––– (1963) *Class in Suburbia,* Englewood Cliffs, N.J.: Prentice-Hall.

Downie, Leonard, Jr. (1974) *Mortgage on America,* New York: Praeger.

Downs, Anthony (1957) *An Economic Theory of Democracy,* New York: Harper & Row.

Erber, Ernest [ed.] (1970) *Urban Planning in Transition,* New York: Grossman.

Fava, Sylvia F. (1956) "Suburbanism as a Way of Life," *American Sociological Review,* 21, (February).

Fischer, C. (1972) "Urbanism as a Way of Life–A Review and an Agenda," *Sociological Methods and Research,* (November), Vol. 1, No. 2,

Friedmann, John (1971) *Retracking America,* New York: Anchor Doubleday.

Frohlich, Norman, Joe Oppenheimer, and Oran Young (1971) *Political Leadership and Collective Goods,* Princeton, N.J.: Princeton Univ. Press.

Gans, Herbert J. (1968) *People and Plans,* New York: Basic Books.

––– (1967) *The Levittowners,* New York: Pantheon.

––– (1962) "Urbanism and Suburbanism as Ways of Life: A Re-evaluation of Definitions," in Arnold Rose (ed.), *Human Behavior and Social Processes,* pp. 625-648.

Gardiner, John A. (1970) *The Politics of Corruption,* New York: Russel Sage.

––– and David J. Olson (1974) *Theft of the City,* Bloomington: Indiana Univ. Press.

Gilbert, Charles (1967) *Governing the Suburbs,* Bloomington: Indiana Univ. Press.

Gitlin, Todd (1969) "Local Pluralism as Theory and Ideology," in Hans Peter Dreitzl, *Recent Sociology,* No. 1, New York: Macmillan.

Goodman, Paul and Percival (1947) *Communitas,* New York: Vintage.

Goodman, Robert (1971) *After the Planners,* New York: Simon & Schuster.

Gottmann, Jean (1961) *Megalopolis: The Urbanized Northeastern Seaboard of the United States,* New York: Twentieth Century Fund.

––– and Robert A. Harper [eds.], (1967) *Metropolis on the Move,* New York: John Wiley.

Greer, Scott (1962) *The Emerging City: Myth and Reality,* New York: Free Press.

––– (1962) *Governing the Metropolis,* New York: John Wiley.

––– (1960) "The Social Structure and Political Process of Suburbia," *American Sociological Review,* 25, (August), pp. 514-526.

Haar, Charles, [ed.], (1972) *The End of Innocence: A Suburban Reader,* Glenview, Ill.: Scott, Foresman.

––– (1959) *Land Use Planning,* Boston: Little, Brown.

Hadden, Jeffrey K. and Louis Masotti, [eds.], (1974) *Suburbia in Transition,* Chicago: Quadrangle.

––– (1973) *The Urbanization of the Suburbs,* Beverly Hills: Sage.

Hawley, Amos, et. al (1975) *Metropolitan America in Contemporary Perspective,* New York: Halsted Press.

Hawley, Willis D. and Frederick M. Wirt (1974) *The Search for Community Power,* Englewood Cliffs, N.J.: Prentice-Hall.

Hook, Sidney [ed.], (1969) *Kenneth Arrow's Welfare Economics: A Symposium,* New York: New York Univ. Press.

James, Franklin J., Jr. and O. D. Windsor (April 1976), "Fiscal Zoning, Fiscal Reform and Exclusionary Land-use Controls," *American Institute of Planners Journal,* pp. 130-141.

Lange, Oskar (1970) *Papers in Economics and Sociology,* Elmsford, N.Y.: Pergamon.

--- (1963) *Political Economy,* Vol. 1, New York: Pergamon.

--- (1938) *On the Economic Theory of Socialism,* Minneapolis: Univ. of Minnesota Press.

Lefebvre, Henri (1971) *Everyday Life in the Modern World,* New York: Harper & Row.

--- (1969) *The Explosion: Marxism and the French Upheaval,* New York: Monthly Review Press.

Long, Norton (1958) "The Local Community as an Ecology of Games," *American Journal of Sociology,* LXIV, No. 3, (November), pp. 251-261.

Lowi, Theodore (1969) *The End of Liberalism,* New York: W. W. Norton.

Lundberg, George, et. al (1934) *Leisure: A Suburban Study,* New York: Columbia Univ. Press.

MacPherson, C. B. (1962) *The Political Theory of Possessive Individualism,* New York: Oxford Univ. Press.

Mannheim, Karl (1951) *Freedom, Power and Democratic Planning,* London: Routledge and Kegan Paul.

--- (1940) *Man and Society in an Age of Reconstruction,* New York: Harcourt, Brace & World.

Olson, Mancur (1971) *The Logic of Collective Action,* Cambridge, Mass.: Harvard Univ. Press.

Paulson, Morton (1972) *The Great Land Hustle,* Chicago: Henry Regnery.

Petersen, William (1966) "On Some Meanings of 'Planning'," *AIP Journal,* (May).

Schnore, Leo F. (1972) *Class and Race in Cities and Suburbs,* Chicago: Markham Press.

--- (1965) *The Urban Scene,* New York: Free Press.

--- and Joy K. O. Jones (1969) "The Evolution of City-Suburban Types in the Course of a Decade," *Urban Affairs Quarterly,* 4 (June), pp. 421-423.

Schon, Donald A., et al. (1976) "Planners in Transition," *AIP Journal,* (April), pp. 193-203.

Scott, Mel (1971) *American City Planning,* Berkeley: Univ. of California Press.

Siegan, Bernard (1972) *Land Use Without Zoning,* Lexington, Mass.: D. C. Heath.

Sobin, L. (1968) *Dynamics of Community Change,* Port Washington, N.Y.: Ira V. Friedman.

Spectorsky, A. C. (1955) *The Exurbanites,* Philadelphia: Lippincott.

Spreiregen, Paul D. [ed.], (1967) *The Modern Metropolis: Selected Essays by Hans Blumenfeld.* Cambridge, Mass.: MIT Press.

Thorns, David (1972) *Suburbia,* London: MacGibbon & Kee.

Whyte, William H., Jr. (1958) "Urban Sprawl," in *The Exploding Metropolis,* by the editors of *Fortune,* Garden City, N.Y.: Doubleday Anchor.

--- (1956) *The Organization Man,* Garden City, N.Y.: Doubleday Anchor.

Wolff, Robert Paul (1969) *The Poverty of Liberalism,* Boston: Beacon.

Wood, Robert C. (1959) *Suburbia–Its People and Their Politics,* Boston: Houghton Mifflin.

--- (1961) *1400 Governments,* Cambridge, Mass.: Harvard Univ. Press.

Zimmer, B. G. (1975) "The Urban Centrifugal Drift," in Amos Hawley, et al., *Metropolitan America in Contemporary Perspective,* New York: Halsted Press, (A Sage Publications book).

Zimmerman, Joseph (1975) "The Patchwork Approach: Adaptive Responses to Increasing Urbanization," in Amos Hawley, et al., *Metropolitan America in Contemporary Perspective,* New York: Halsted Press, (A Sage Publications book).

Zschock, Dieter K. [ed.], (1968) *Brookhaven in Transition: Studies in Town Planning Issues,* Stony Brook, N.Y.: SUNY at Stony Brook.

ABOUT THE AUTHOR

MARK GOTTDIENER holds graduate degrees in Economics and Sociology and has worked as a regional planner in New York and New Jersey. He has taught Economics at the University of Minnesota and at Queensboro Community College and Sociology at Dowling College and at SUNY at Stony Brook. Currently, he is Assistant Professor of Sociology at Brooklyn College and Acting Chairperson of the Interdisciplinary Program in Urban Studies.

SAGE LIBRARY OF SOCIAL RESEARCH

1. **DAVID CAPLOVITZ:** The Merchants of Harlem
2. **JAMES N. ROSENAU:** International Studies and the Social Sciences
3. **DOUGLAS E. ASHFORD:** Ideology and Participation
4. **PATRICK J. McGOWAN and HOWARD B. SHAPIRO:** The Comparative Study of Foreign Policy
5. **GEORGE A. MALE:** The Struggle for Power
6. **RAYMOND TANTER:** Modelling and Managing International Conflicts
7. **ANTHONY JAMES CATANESE:** Planners and Local Politics
8. **JAMES RUSSELL PRESCOTT:** Economic Aspects of Public Housing
9. **F. PARKINSON:** Latin America, the Cold War, and the World Powers, 1945-1973
10. **ROBERT G. SMITH:** Ad Hoc Governments
11. **RONALD GALLIMORE, JOAN WHITEHORN BOGGS, and CATHIE JORDAN:** Culture, Behavior and Education
12. **HOWARD W. HALLMAN:** Neighborhood Government in a Metropolitan Setting
13. **RICHARD J. GELLES:** The Violent Home
14. **JERRY L. WEAVER:** Conflict and Control in Health Care Administration
15. **GEBHARD LUDWIG SCHWEIGLER:** National Consciousness in Divided Germany
16. **JAMES T. CAREY:** Sociology and Public Affairs
17. **EDWARD W. LEHMAN:** Coordinating Health Care
18. **CHARLES G. BELL and CHARLES M. PRICE:** The First Term
19. **CLAYTON P. ALDERFER and L. DAVE BROWN:** Learning from Changing
20. **L. EDWARD WELLS and GERALD MARWELL:** Self-Esteem
21. **ROBERT S. ROBINS:** Political Institutionalization and the Integration of Elites
22. **WILLIAM R. SCHONFELD:** Obedience and Revolt
23. **WILLIAM C. McCREADY and ANDREW M. GREELEY:** The Ultimate Values of the American Population
24. **F. IVAN NYE:** Role Structure and Analysis of the Family
25. **PAUL WEHR and MICHAEL WASHBURN:** Peace and World Order Systems
26. **PATRICIA R. STEWART:** Children in Distress
27. **JUERGEN DEDRING:** Recent Advances in Peace and Conflict Research
28. **MOSHE M. CZUDNOWSKI:** Comparing Political Behavior
29. **JACK D. DOUGLAS:** Investigative Social Research
30. **MICHAEL STOHL:** War and Domestic Political Violence
31. **NANCY E. WILLIAMSON:** Sons or Daughters
32. **WERNER LEVI:** Law and Politics in the International Society
33. **DAVID L. ALTHEIDE:** Creating Reality
34. **ALLAN LERNER:** The Politics of Decision-Making
35. **PHILIP E. CONVERSE:** The Dynamics of Party Support
36. **CHARLES L. NEWMAN and BARBARA R. PRICE:** Jails and Drug Treatment
37. **CLARENCE L. ABERCROMBIE III:** The Military Chaplain
38. **MARK GOTTDIENER:** Planned Sprawl
39. **ROBERT L. LINEBERRY:** Equality and Urban Policy
40. **PATRICK M. MORGAN:** Deterrence
41. **VLADIMIR A. LEFEBVRE:** The Structure of Awareness
42. **ANDREA FONTANA:** The Last Frontier